ADAPTIVE USER SUPPORT

**Ergonomic Design of Manually and
Automatically Adaptable Software**

COMPUTERS, COGNITION, AND WORK

A series edited by:
Gary M. Olson, Judith S. Olson, and Bill Curtis

ADAPTIVE USER SUPPORT

Ergonomic Design of Manually and Automatically Adaptable Software

Edited by

Reinhard Oppermann

Institute for Applied Information Technology
GMD, German National Research Center for Computer Science

CRC Press
Taylor & Francis Group
Boca Raton London New York

CRC Press is an imprint of the
Taylor & Francis Group, an **informa** business

First Published by
Lawrence Erlbaum Associates, Inc., Publishers
10 Industrial Avenue
Mahwah, New Jersey 07430

Transferred to Digital Printing 2009 by CRC Press
6000 Broken Sound Parkway, NW Suite 300, Boca Raton, FL 33487
270 Madison Avenue New York, NY 10016
2 Park Square, Milton Park Abingdon, Oxon OX14 4RN, UK

Library of Congress Cataloging-in-Publication Data

Adaptive user support : Ergonomic design of manually and
  automatically adaptable software / edited by Reinhard Oppermann.
      p. cm. -- (Computers, cognition, and work)
  Includes bibliographical references and index.
  ISBN 0-8058-1655-0
  1. Human-computer interaction. 2. User interfaces (Computer
systems)   I. Oppermann, Reinhard, 1946-   . II. Series.
QA76.9.H85A33 1994
005.1'01'9--dc20   94-21630
                                               CIP

Publisher's Note
The publisher has gone to great lengths to ensure the quality of this reprint
but points out that some imperfections in the original may be apparent.

Contents

Introduction
Reinhard Oppermann

Chapter 1:   Adaptability: User-Initiated Individualization
Reinhard Oppermann and Helmut Simm

Chapter 2:   Adaptivity:
System-Initiated Individualization

Mette Krogsæter and Christoph Thomas

Chapter 3:   A User Interface
Integrating Adaptability and Adaptivity

Mette Krogsæter, Reinhard Oppermann, and Christoph Thomas

Chapter 4:   HyPLAN :
A Context-Sensitive Hypermedia Help System

Thorsten Fox, Gernoth Grunst, and Klaus-Jürgen Quast

Chapter 5: Configurative Technology: Adaptation to Social Systems Dynamism

Michael Paetau

List of Contributors

Thorsten Fox GMD.FIT, P.O.B 1316, D-53731 Sankt Augustin, Germany

Gernoth Grunst GMD.FIT, P.O.B 1316, D-53731 Sankt Augustin, Germany

Mette Krogsaeter Oslo College of Engineering, Cort Adelers gate 30, N-0254 Oslo, Norway

Reinhard Oppermann GMD.FIT, P.O.B 1316, D-53731 Sankt Augustin, Germany

Michael Paetau GMD.FIT, P.O.B 1316, D-53731 Sankt Augustin, Germany

Klaus-Jürgen Quast GMD.FIT, P.O.B 1316, D-53731 Sankt Augustin, Germany

Helmut Simm GMD.FIT, P.O.B 1316, D-53731 Sankt Augustin, Germany

Christoph Thomas GMD.FIT, P.O.B 1316, D-53731 Sankt Augustin, Germany

Introduction[1]

Reinhard Oppermann

> Probably the most important
> fundamental that is being ignored
> by business today is staying close
> to the customer.
>
> *Alberto Giacometti*

This volume presents the result of the SAGA project about user support. SAGA is a German acronym for "software-ergonomic analysis and design of adaptivity". The main attention in the project was addressed to the support a system can generate in evaluating the user's interaction with the system. The approach was demand driven on the basis of interaction analyses. The first point of emphasis was to provide flexible adaptation possibilities so the user could adapt insufficient system elements. The second point was a context-sensitive help environment to explain complex or complicated system components.

Computers dominate the working environment of many people. Application systems are not designed for a particular user and a particular task. They are designed and distributed for a class of users and a set of tasks. This is true for both standard systems and specialized systems. The individual users change, and the tasks vary and evolve. In the last few years, much work has been done in making systems more flexible. Flexibility is needed not only in the design phase, but also during usage. It makes the user more independent of the designer, and does not force the designer to decide about user-specific optima before the user works with the system.

---

[1] The investigation reported in this volume was part of the research program "Assisting Computer" (AC) of the Institute for Applied Information Technology of the GMD (the German National Research Center for Computer Science). The purpose of the AC program is to investigate new kinds of support systems for computer applications exploiting knowledge about the systems, the tasks, and the user characteristics and habits (Hoschka 1991). The project was supported by a grant from the "Work and Technology" fund of the Ministry of Science and Technology of the Federal Republic of Germany under grant No. 01HK537/0.

A flexible system should give the user greater freedom; improve the correspondence between user, task and system characteristics; and increase the user's efficiency. The goal is to fit the system to a specific user and to a specific task not only on the basis of requirements analysis made at the design phase, also by prolonging the flexibility of the system into the usage phase by its adaptation capability.

One way to obtain flexibility is to provide multiple options for functionally equivalent interaction styles. This approach increases the user's freedom, but it also increases the complexity of the system. Another kind of flexible system is a system that can be tailored. Two variations are known in the literature: adaptable and adaptive systems. A system is called *adaptable* if it provides the end user with tools that make it possible to change the system characteristics (see the chapter by Oppermann and Simm for a detailed description of adaptable systems). It is an attractive objective and a current endeavor of commercial software production to provide the user with facilities for tailoring the system according to his personal tasks and needs. This kind of individualization gives control over the adaptation to the user. It is up to the user to initiate the adaptation and to employ it. There is a wide spectrum of tools and methods in commercial applications for customizing the interface of the system. For example, special menus or key combinations open up access to interface presentation, naming, and dynamics; there are macro mechanisms to combine functions. The facilities can be used by the "end-user", by a "superuser", or by a technical specialist (see Edmonds 1981).

A system is called adaptive if it is able to change its own characteristics automatically according to the user's needs (see the chapters by Krogsæter, Oppermann, and Thomas and by Fox, Grunst, and Quast for examples of adaptive systems also see the chapter by Krogsæter and Thomas for a general discussion of adaptive systems). The self-adapting system is the most common conception of adaptivity. Modification of interface presentation or system behavior depends on the way the user interacts with the system. The system initiates and performs changes appropriate to the user, his tasks, and specific demands.

The discussion about adaptivity does not show a clear superiority of one solution over the other. There is empirical evidence that users have difficulties in using adaptable features of systems or, at any rate, don't use them. On the other hand, self-adapting systems include other inherent problems of recording user interaction styles and of taking control over decisions about the best way to perform a task. In this book, we explore the potential of both types of systems, in terms of actually existing applications and through the presentation of two prototypes that we developed ourselves. One of these two prototypes combines adaptability and adaptivity by offering an environment with shared initiative, shared decision making, and shared execution of adaptations between the user and the system.

1. Goals and Problems of Adaptive Systems

In order to find ways to make man machine processes more productive, current research investigates the feasibility of a new kind of application that will be designed to act as an assistant to the user rather than as a rigid tool. An assistant exhibits different behavior than do current rigid systems. An ideal human assistant adopts the perspective of the client: The assistant learns about the client and knows about the client's needs, preferences, and intentions. In human-human communication, both participants can adapt their behavior according to the characteristics of the partner. There are many initial clues about the character of the partner, the intentions to open a communication, and the style of communication. The communication process enriches and refines the knowledge of both partners about each other: They learn from the verbal and nonverbal behavior. This knowledge enables the partners to improve the communication process to accelerate the discovery of common or conflicting aims, to optimize communication, and thereby to increase their satisfaction with the interaction. So much for human-human communication—what about human computer communication? What can a technical assistant adopt from a human assistant? Why not transfer the effective mechanism of human-human communication to the interaction between a user and a system? Why not let the system learn about the user and model the characteristics of the user in its knowledge-base, in order to adapt itself to the user? There has been some research in the last 10 years on developing methods for building user models and adaptive features. These attempts did not, however, result in any final conclusions—they show, at best, anecdotal impressions. In this volume, we will discuss problems and possibilities of adaptive systems and present examples of unpretentious but successful adaptation features.

Adaptivity in the form of an adaptive system is based on the assumption that the system is able to adapt itself to the wishes and tasks of the user by an evaluation of user behavior, thus breaking down a communication barrier between man and machine (see Hayes, Ball, & Reddy 1981). Consequently, adaptivity is basically implemented by referring actions to action patterns and finally to action plans on which the actions of the person are based. These action patterns must be identified in the evaluation of user action steps. The interpretation of user actions serves to identify user intentions. Deducing user intentions from user actions may be done in several ways. For example, having deduced an action plan, the system may initiate a dialogue to identify the user's real intentions. The system presents alternative interpretations, and the user selects the appropriate one. In this case, the current action is checked for specific regularities agreeing with previous action sequences. Where agreements are detected, appropriate subsequent actions will be assumed. In this way, computer systems should become able to adapt to anticipated user behavior in a way analogous to human-human communication.

The goal of adaptive systems is to increase the suitability of the system for specific tasks; facilitate handling the system for specific users, and so enhance user productivity; optimize workloads, and increase user satisfaction. The ambition of adaptivity is that not only that "everyone should be computer literate", but also that "computers should be user literate" (Browne, Totterdell, & Norman 1990). There are examples of adaptive systems that support the user in the learning and training phase by introducing the user into the system operation. Others draw the user's attention to tools he is not familiar with to perform specific operations for routine tasks more rapidly. Evaluation of system use is designed to reduce system complexity for the user. In the event of errors or disorientation on the part of the user, or where the user requires help, the adaptive system is to provide task- related and user-related explanations. Automatic error correction is to be employed when user errors can be uniquely identified. The user is spared the necessity of correcting obvious errors.

An adaptive system is based on knowledge about the system, its interface, the task domain, and the user (see Norcio & Stanley 1989). It must be able to match particular system responses with particular usage profiles. Edmonds (1987) described five areas the system can take into account: user errors, user characteristics, user performance, user goals, and the information environment. He suggested that user errors are the most prominent candidate for automatic adaptations. Benyon and Murray (1988) proposed adaptations of the system to the environment of the user at the interface level rather than at the user's conceptual task level. Salvendy (1991) referred to personality traits, such as field dependence versus field independence (see also Fowler, Macaulay, & Siripoksup 1987), or introversion versus extroversion.

There are attempts to find psychological theories to describe human characteristics that are more or less persistent and are of more or less relevance for the adaptation process (see Benyon & Murray 1988; Benyon, Murray, & Jennings 1990; Norcio & Stanley 1989; Veer et al. 1985). To the best of our knowledge, no working applications have been developed whose behavior depends on the relationship between a personal trait and system behavior. The most solid candidate for a relationship between personal behavior (although without an underlying general personality trait) and system behavior is a system reported in Browne, Totterdell, and Norman (1990). The adaptive feature of the system is concerned with the correction of misspellings. The users can be described by their error frequency (missing letter, extra letter, wrong letter, transposed adjacent letters) and be supported by appropriately arranged lists of suggestions for correction. The limitation of this example is obvious. It cannot be generalized and the advantage is limited to a reduction of some milliseconds in the correction time. The same kind of limited advantage (but without the disadvantage of dynamic menus as in the next example) is provided by adaptive menu defaults, where the mouse cursor is prepositioned at the user's most likely selection (see Browne, Totterdell, & Norman 1990). In another example, the adaptive effect is a reordering of menus, where the most frequently used menu options are presented at the top. This leads to a decreased access time but also—especially in the course of learning the system—to an increased search time (see Mitchell & Shneiderman 1989). Proofs for the feasibility and accountability of adaptive systems with respect to personality traits, such as introversion/extroversion, field dependence/field independence (see Salvendy 1991), where relevant system behavior corresponds to personality differences, have yet to be shown. Considering the difficulties in developing adaptive systems that respond to specific personality traits, we concentrate on task-specific regularities that could be related to differentiated system behavior.

There is an ongoing discussion about the possible dangers of adaptivity, in the form of drawbacks and problems that are in opposition to the objectives of adaptive systems. The user is observed by the system, actions are recorded, and information about the user and his task is inferred. This gives rise to data- and privacy-protection problems (i.e., the individual user is exposed to social control). This social monitoring possibility is aggravated by monitoring of the user by the technical system. The user is, or feels, at the mercy of the control and dynamics of the system. A system hidden from the user, takes the lead, makes proposals, and, in some cases, gives help and advice without being asked. This makes the technical system appear presumptuous and anthropomorphized. Another problem with self-adapting systems is that the user is exposed to the pressure of adapting to the developer's conceptual model, on which the system is based. Finally, the user can be distracted from the task by following modifications of the system or modification suggestions. The user's overall grasp of the system structure and system capabilities is lost and transferred to the system. The user's mental model of the system, which has been acquired in using the system, becomes confused.

In order to overcome at least some of the specified problems of adaptive systems, while still achieving their objectives, the majority of authors propose that the user control how the system adapts to user behavior. This can be done by (a) providing means for the user to activate and deactivate adaptation for the overall system or individual parts of the system before any adaptation is made or after a specific adaptation state has been reached; (b) offering the adaptation to the user in the form of a proposal that he can accept or reject, or enabling him to select among various possibilities of adaptation modification;

(c) enabling the user himself to define, in an adaptation-resistant manner, specific parameters required for adaptation by the system; (d) giving the user information on the effects of the adaptation modification, which may protect the user from surprises; and (e) giving the user sole control over the use of his behavior records and their evaluation.

2. Elements of Adaptive Systems

Adaptive systems (i.e., their architecture, their behavior, their application and their limitations) can generally be described as consisting of three parts: an afferential, an inferential, and an efferential component. This nomenclature borrows from a simple physiological model of an organism with an afferential subsystem of sensors and nerves for internal and external stimuli, with an inferential subsystem of processors to interpret the incoming information, and with an efferential subsystem of effectors and muscles to respond to the stimuli. Unlike mechanical systems, a biological system is capable of reflecting on the input by considering the meaning and the context of sequences, rather than just reacting to an input pattern with a conditioned response or a monosynaptic reflex.

Afferential Component of Adaptivity. Adaptive systems observe and record user behavior and system reactions, in addition to directly executing commands according to their input. They recognize regular interaction patterns, frequently occurring errors, ineffective courses, and so on. Adaptive systems can gather data on different levels ranging from, for example, information about key strokes, mouse movements, and clicks, through information about functions or error types, on up to task-specific information.

Inferential Component of Adaptivity. Adaptive systems analyze the gathered data in order to draw conclusions (i.e., to identify from user behavior possible indicators for adaptation). This is the most crucial point. The inferential component decides in which way the system should modify its behavior in order to correspond to the actual usage profile. This implies that a basis must be specified (a theory, a set of rules) for drawing inferences. This also implies that the kind of data to be recorded (afferential component) and how the system should be adapted (efferential component) must be defined. Thus, the inferential component is the switchbox of an adaptive system. The inference of adaptations can be based on rules and heuristics that are represented by a model of appropriate relationships between the system and the user. The adaptive system can describe the system characteristics in a system model, the user characteristics in a user model, and the task characteristics in a task model (see Norcio & Stanley 1989).

Efferential Component of Adaptivity. Adaptations lead to modifications of the system's behavior. The change may concern presentation of objects, functions, or tools; default values for parameters; sequences of dialogues, or system messages (status, execution, help, error messages). Changes may occur in both the application itself and the error handling and help system. The effect of the adaptation can also be differentiated by the degree of determinacy of the changes: They can result in one particular effect being offered by the adaptive system or in a proposed range of alternatives to be selected from by the user.

3. Determinism and Autonomy

It is difficult for the inferential component to be completely deterministic when deciding on modifications. To infer a specific outcome from a specific usage profile in a strict if then else relationship overburdens the inferential component when possibly unforeseeable contexts influence the value of an adaptation. The user may only accept or reject the adaptive decision. On the other hand, a range of alternatives with different views on the system is a less problematic alternative. Such proposals are less ambitious and give more freedom to the user. The user can make use of an introduced set of alternatives and can navigate through a space of proposed adaptations. We see in the chapters that follow two classes of adaptations—by Krogsæter, Oppermann, and Thomas and by Fox, Grunst, and Quast—how the demand for a help and adaptation space rather than the presentation of one-shot solutions was theoretically and empirically identified.

The concept of adaptive systems raises the more general question of autonomy. Machines are perceived as technical artifacts that are designed to transform a given input into a particular output. The transformation is unequivocally describable. This is called a *trivial machine* (according to the system theory of Heinz von Foerster, see Sprengler 1992; Ulrich & Probst 1988). Trivial machines can be complicated, but they are not perceived of as complex.[2] Complex systems are biological or social systems. Complex systems are creative. They not only react to stimuli but also reply. They learn and are able to acquire a new action repertory. Technical systems are not creative. Changes from expected behavior are perceived as failure. Is an adaptive system a trivial machine? Is it creative? No, it is not creative. The designer is creative, and the adaptive system supports the creativity of the user. Adaptive systems behave deterministically. They follow explicit rules that are designed by the constructor. Their behavior can be predicted. They differ from classical ma-

---

2 *Complexity* is perceived as a strict term in this context, but is used in a broader sense in the remainder of this volume.

chines in their dependence on an adaptive reaction on the system's state. The reaction takes history into account and it considers the dimension of time. Adaptive systems are complex systems in that they are able to adopt a great number of states that moderate the system's behavior in a given situation so that this behavior appears to be creative, but actually is not. It differs from the behavior in an analogously perceived situation with the same input but different state conditions. The limitations of present and principal machine capabilities should be kept in mind when talking about adaptive systems.

4 . Demand-Driven Development of Adaptive Systems

Our experience with attempts to implement adaptivity (our own and others in the literature) showed that the development of adaptive systems should not be primarily based on technological capabilities. Rather, solutions should reflect demand-driven design concentrating on the users' needs. The designs of the two systems presented in this volume, a context-sensitive help environment and an adaptively supported adaptation environment, were not strictly specified in advance. They started with an analysis of requirements for help and adaptation support in a given application for a set of realistic tasks. The selection and presentation of help and adaptation facilities reflected the results of design–evaluation–redesign cycles. Recognizing the limited functionality of existing examples of adaptive systems in ad hoc prototypes, we started with a complex commercial application with an advanced user interface (the spreadsheet program EXCEL™). This program offered opportunities for afferential observations and efferential interventions. The afferential opportunity was a record of the user's actions as the basis for making inferences about user needs from user actions. The efferential opportunity was the possibility to build a customized application consisting of a set of functions implemented by macros of the EXCEL system.

5 . Contents of the Volume

The chapter "Adaptability: User-Initiated Individualization" by Simm and Oppermann describes and discusses the current state of the art of adaptability. Many commercial software products may be provided with a considerable degree of adaptability. No consistent concept of tools and methods has been employed by designers of adaptation facilities. This makes it difficult for the user to detect and use the potential of adaptability. Users report difficulties making adaptations for actual tasks.

The chapter "Adaptivity: System-Initiated Individualization" by Krogsæter and Thomas gives an overview of work that has been done on adaptive systems from the user, as well as the software engineering, point of view. One type of assistance is provided by help systems that adjust the help given to the user depending on the task context or user traits, or both. Other systems change or offer to change the user interface or system functionality in order to better suit individual users' needs. In order to implement adaptive features, various knowledge-bases (or models) must be included in the systems. A user-modeling component attempts to determine characteristics and preferences of individual users, whereas a task-modeling component tries to figure out the tasks the user is trying to achieve. For modeling and for mapping model states to appropriate system responses, techniques for the design and implementation of knowledge-based systems are usually applied.

The chapter "A User Interface Integrating Adaptability and Adaptivity" by Krogsæter, Oppermann, and Thomas describes a prototype of a system that incorporates adaptive and adaptable capabilities to support users in typical spreadsheet operations. The system provides the user with adaptations of the interface: Defaults, menu entries, and key shortcuts are subjects of adaptation.[3] Two kinds of adaptation facilities were developed. The first was a tool that the user could apply for adapting the system. The second was adaptive, with the system suggesting an adaptation to the user. These two components work together and allow the user to introduce adaptations himself and to accept or reject proposed adaptations. It is seen as important that users have the opportunity to make their own adaptations with the adaptation tool. The adaptive component may only present suggestions proposing modifications of the interaction appropriate for the actual use of the system. The user may select from offered suggestions. The user's control includes two aspects. The first is the opportunity to work autonomously without having to wait for, or being dependent on, the adaptive component. The second is that the user has the authority to accept or reject adaptations suggested by the system. The introduction of an adaptation always requires an explicit decision by the user.

Adaptations are either suggested by the system or introduced by the user while performing a task. The resulting adaptation may be manifested by modified defaults, additional menu entries, or key shortcuts that are integrated in the familiar user interface. The controversially discussed location of initiative for adaptations is split between the system and the user, rather than being placed exclusively with the user or with the system. The adaptations can be

---

[3] An adaptation of the functionality of an application is not considered in this system. To provide the user with the most appropriate functions for a task requires more knowledge about the user and the task than can be acquired by analyzing the user's actual usage record.

prepared by the system, but the user has the opportunity to employ them and to specify values and names. The development process of this prototype, as well as the evaluation of the resulting design in user experiments, are described in that chapter.

The chapter "HyPLAN: A Context-Sensitive Adaptive Hypermedia Help System" by Fox, Grunst & Quast introduces a help and tutorial system for Excel. As described by Edmonds (1987), automatic adaptations can be applied most successfully to error support. Indeed, most examples of adaptive systems support the user in error or problem situations. The development of HyPLAN puts into practice ideas of adaptivity concerning the design and use of help systems. At the beginning of the development, we studied interactions between human tutors and users of Excel working on tasks typical of those in their working domain. We tried to transform the observed success models of human interaction and users' demands in the design of the help system. Resulting prototypes were tested in the laboratory, and the results of these were used for a new design cycle. In this sense, HyPLAN is adapted to typical human cognitive demands and abilities. On the other hand, HyPLAN is an adaptive system because help access is prepared according to the recent use of Excel. By hitting a Help button, the user is offered a pertinent selection of help topics. The topics are selected by a plan recognizer matching the user's actions with known action plans represented in a knowledge-base. Mapping plan executions to typical usage problems and plans representing suboptimal usage of Excel, the selection of help topics tries to orient the user in his search for efficient solutions. The first section of the chapter describes the analytic methodology driving the design, recording, and analyses of the empirical experiments. The second section introduces the derived hypermedia interface and central concept of the help and tutorial system. The last section focuses on the plan recognition component.

In the final chapter, "Configurative Technology: Adaptation to Social Systems Dynamism", Paetau presents the sociological part of our investigations. Sociological studies, in particular, have suffered from the problem that questions about the potential consequences of a particular technology cannot be answered away from the socio-organizational context; and it is precisely this context that so often cannot be anticipated, at least not adequately, in the early phases of technological development. In order to realize a "prospective design requirement" Paetau pleads for an analysis of visions, models, and concepts. With this approach, he takes the existing knowledge of a pioneering technology that is currently being developed and relates it to evolving socio-organizational constellations.

The most important finding of the sociological approach is that the discussion, which has already been going on for quite some time, of how to adapt technology to humans must break free of some familiar ideas inherited from traditional ergonomics. Whereas it was still possible for traditional machine-oriented ergonomics to, as it were, measure humans and their work and derive ideal–typical constructions and norms by which individual machine elements could be constructed, this is no longer the case in the design of computer systems. Computers are a means of mechanizing mental activities and socio-organizational structures. These elements are subject to substantially stronger dynamics of change than, say, the parameters of man's physical features, such as the length of an average arm, sitting postures, and so on. In the development of human mental work, we are dealing with an evolutionary process whose range and speed cannot be compared to the evolution of the physical characteristics of human work. As a result of this finding, Paetau describes the concept of *configurativity*, which has consequences for the problem of the orientation of adaptation and modification features, for the initiator of adaptations (i.e., the one who carries out the system adaptation measures), and the means by which system adaptations are realized.

The aim of this volume is to present an overall perspective of adaptations performed by shared competence of the user and the system, to show how to base the development of prototypes on cycles of design, and to discuss the presented concept in technical, psychological, and sociological perspectives.

References

Benyon, D., & Murray, D. (1988):
Experience with adaptive interfaces. *The Computer Journal, 31,* 465-473.

Benyon, D., Murray, D., & Jennings, F. (1990):
An adaptive system developer's tool-kit. In: D. Diaper et al. (Eds.), *Human–computer interaction–INTERACT '90.* Amsterdam: Elsevier Science Publishers, pp. 573-577.

Browne, D., Totterdell, P., & Norman, M. (1990):
Adaptive user interfaces. London: Academic Press.

Edmonds, E. A. (1981):
Adaptive man–computer interfaces. In: M.C. Coombs & J.L. Alty (Eds.), *Computing skills and the user interface.* London: Academic Press, pp. 389-426.

Edmonds, E. A. (1987):
Adaptation, response and knowledge. *Knowledge- Based Systems, 1 ,* 3-10.

Fowler, C.J.H., Macaulay, L.A., & Siripoksup, S. (1987):
An evaluation of the effectiveness of the adaptive interface module (AIM) in matching dialogues to users. In: D. Diaper & R. Winder (Eds.), *People and computers III.* Cambridge: Cambridge University Press, pp. 346–359.

Hayes, P., Ball, E., & Reddy, R. (1981):
Breaking the man–machine communication barrier. *IEEE Computer, 14,* 3-30.

Hoschka, P. (1991):
Assisting computer: A new generation of support systems. *Proceedings of the 4. International GI-Congress "Knowledge Based-Systems— Distributed Artificial Intelligence",* Berlin: Springer.

Mitchell, J., & Shneiderman, B. (1989):
Dynamic versus static menus: An exploratory comparison. *SIGCHI Bulletin, 20,* 33-37.

Norcio, A. F., & Stanley. J. (1989):
Adaptive human–computer interfaces: A literature survey and perspective. *IEEE Transactions on Systems, Man, and Cybernetics, 19,* 399-408.

Salvendy, G. (1991):
Design of adaptive interfaces and flexible mass production of knowledge-based systems. In: H.-J. Bullinger (Ed.), *Human aspects in computing: Design and use of interactive systems and work with terminals.* Amsterdam: Elsevier Science Publishers, pp. 55-68.

Sprengler, R. K. (1992):
   Mythos Motivation. Frankfurt: Campus.

Ulrich, H., & Probst, G.J.D. (1988).
   Anleitung zum ganzheitlichen Denken und Handeln. Bern: Haupt.

Veer, G. van der, Tauber, M., Waern, Y., & Muylwijk, B. van (1985):
   On the interaction between system and user characteristics. *Behaviour
   & Information Technology, 4,* 289-308.

Chapter 1
Adaptability: User-Initiated Individualization

Reinhard Oppermann and Helmut Simm

1. Introduction

Systems are flexible when they allow users to design procedures and display results according to individual preferences. Various authors (Ackermann & Ulich 1987, pp. 131ff.; Grob 1985; Triebe 1980; Ulich 1978; Zülch & Starringer 1984) have pointed out that there can be no single "best way" for workflows or activities. It has also been shown that interindividual differences in procedures are possible without an inevitable loss of efficiency (Triebe 1980). Conversely, strict insistence on specific procedures may only mean that inefficient methods and suboptimal solutions are being cemented. The main practical and theoretical foundation for individualization is the principle of differential and dynamic work design developed by scientists at the ETH Zürich (see Ulich 1978) on the basis of occupational and organizational psychology.

A question central to the exploitation of the potential of flexibility is whether employees who are strictly task-oriented are ready to consider using a tool. Or are they interested in taking a mental look at their equipment to such an extent that they are able to customize systems to suit the needs of their task and their individual or group preferences, such that their application systems would assume the character of open tools. The preconditions for such customization, of course, are a certain level of knowledge, professional competence and the qualifications to match, and the availability of enough time, which together generate the necessary design potential. From the employer's standpoint, the question is whether the provision of such time is necessary or whether system users can find time on their own.

Flexibility has no value in and of itself. Users of flexible or adaptable systems may feel that these options are an enrichment, motivating them and increasing their willingness to make changes, now that they are able, on their own, to modify the system in ways they feel are sensible, and now that they, as users, are in charge. But users also may feel that existing adaptation options are an extra strain that only keeps them from their real work. The inclination to adapt and reflect at a higher cognitive level certainly depends not only on interindividual differences, but equally on the way individuals are embedded in the overall work environment. For the sake of humane workplace design, what we need is the creation of systems in which the human is the "master" (Kubicek 1979, pp. 25f.), that is, systems that support the user's work and broaden his scope. Systems in which the user is the "slave" and the system monitors the user and assumes control, systematically restricting his room to maneuver and breaking his work down into segments so that he loses sight of the overall context, are not considered helpful.

So, if account is to be taken of such findings about interindividual differences within a work system, we must consider the principle of flexible work design (Ulich 1987, pp. 86f.), while bearing in mind that this is only one general form of the principle of differential work design, which considers "the range of different work systems available at one time between which the employee can choose" (Ulich 1978, p. 568), ensuring optimal development of personal potential. If personality is perceived as a potential under individual development, however, we must consider interindividual differences. The principle of differential work design must be supplemented by the principle of dynamic work design, meaning the "extension of existing or the creation of new work systems as well as the possibility of switching between different work systems" (Ulich 1978, p. 568). Note that necessary, but not sufficient, control aspects, both for the process and the result of the work design and for the user-controlled dialogue, are transparency and predictability (Spinas, Troy, & Ulich 1983, p. 17). Indeed, the ability to be influenced, which presupposes transparency and predictability, but goes beyond them, is a crucial criterion for ensuring control functions.

It is evident that there is a wide scope for considering the aforementioned principles and criteria precisely in the area of computer-assisted activities (office work, manufacturing, etc.). In particular, interindividual and intraindividual variations over time in the different degrees of users' experience and familiarity with software call for suitable man–computer interfaces. This is what makes flexible/individualized and user-controlled system design crucially important (Ulich 1987, pp. 87ff.). In this case, rigid systems appear inappropriate, because control then lies with the computer, and the user has no options in the way of selection, influence, or control.

Today, software cannot be, and is not being, written for single users, but always for larger groups. Developers generally try to form user models and homogeneous user subgroups that can be described by various similar features. Common classifications are breakdowns into *beginners*, *occasional users*, and *experts* (Krause 1988, pp. 6ff.) or into *naive*, *occasional*, and *practiced users* (Hoffmann 1988, pp. 30f.). Such user classifications made for the sake of system design are much too rough, however, because the averaging of user features within even small groups—although this does help avoid more serious distortions–still leaves huge differences between the individual and the "average user".

Anyway, the individual user is not always very uniform either: He may be very familiar with some applications, making him an expert in such cases, and quite unfamiliar with others, in which he is a beginner; he can learn, but also forget, and he may, in an extreme case, even forget his own modifications

(Zapf 1990). This being so, it is obvious that there can be no "best" or "optimal" user interface (Rathke 1987, p. 122), so that the user can best benefit from flexible systems.

The goal of flexibility can be reached in various ways. The system may provide various access options that are functionally equivalent, allowing the user to choose between them. We call this *variety*. Or, flexibility may come in the form of options for *individualization* by preparatory activities. Such options may, in their turn, be implemented in two ways: in the form of active system changes made by the user (adaptability) or in the form of system changes made by the system (auto-adaptivity). Traditionally, a strict distinction has been made between these alternatives. A system is said to be *adaptive* or *self-adapting* if it completes an adaptation on its own initiative based on its evaluation of user actions. A system is said to be *adaptable* if it can be modified by the user to suit his needs: The system provides the tools for the user to make the adaptation. The problems involved in this dichotomy are discussed in the introduction to this volume; while the chapter by Krogsæter, Oppermann, and Thomas shows how the two principles can be combined.

Variety in the available handling options is certainly one useful way to take into account differences between users and between tasks, because this enables the user to decide at any time, ad hoc, about the way he wants to use the system, for example, whether he wishes to call a function via menu or key combination. It is not generally necessary that he learn multiple ways, but he can start using an alternative whenever he wants. Viewed this way, variety is certainly a royal road to flexible system use, but it is a road studded with pitfalls. A system marked by variety tends to become unmanageable in size and may become too complex. The user no longer has a command of the full range of interactive options, the other two routes to flexible system use also have a role to play: individualization via auto-adaptivity and adaptability. This chapter deals with adaptability. We start with the options offered by commercial systems, examining the features open to adaptation and the methods of adaptation employed. The final section presents and discusses published findings on how users take advantage of adaptability.

If we are to recognize and explore not only potential benefits, but also the problematic aspects of adaptability, it is not enough to ask users whether, why–and, possibly, which–adaptations have been made. This may be necessary, but is by no means sufficient. In any discussion with users, the analyst must also be able to recognize which adaptation options were *not* used in a particular situation. The user must be asked this, as well. It is also useful and interesting to discuss alternatives and variants, so the observer must have precise knowledge of the adaptation options offered by specific software. Of course, isolated findings are not enough either: Any comparative view claiming to be objective requires that particular cases be considered within a frame of reference that defines the relevant scale and forms of adaptation. Such an instrument also makes it much easier to assess any changes made by the user and to check whether perhaps they merely mitigate weaknesses in the software (so that they are not genuine adaptations to tasks or to user preferences). Only part of such a frame of reference can be derived by deductive methods, namely, its rough outline. Completion of the fine structure requires that we vet and classify our practical findings. The actual value of such a frame of reference depends on how naturally and completely it is able to structure these empirical findings.

The following sections present such a frame of reference for adaptations. It considers adaptations that can be made either by the user himself or by some other person in the use environment without recourse to the manufacturer's expertise. In its empirical content, it is the result of scrutinizing and following up developments in some familiar nonspecialized office applications for personal computers (the Macintosh line, in particular) with the object–suggested earlier as a question–of systematically recording the available adaptation options. We were mainly concerned with word processing and spreadsheet software, but also considered operating systems or their extensions and some relevant utilities. As for the products themselves, the wide range of software available made an exhaustive treatment impossible, so this was not our intention. As was said earlier, our chief object was to use familiar programs to analyze the adaptability options available today and to structure them for the sake of comparisons. Because such an approach tends to be abstract, our purpose was not to give a comprehensive account of individual products and their adaptability, but to illustrate each dimension of adaptability by referring to the characteristic features of one or more products.

2. Areas of Adaptability

This overview of the possible areas for making adaptations, that is "what" can be adapted, is divided into two main topics: adaptations of functionality and interface adaptations. The first topic covers options for adapting the range and behavior of the features of a system, and includes modifying the scope of functionality, employing defaults, and setting up trigger options. The second topic covers options for modifying the access to these features, their interactive dynamics, and their use of screen layout.

2.1. Adaptation of Functionality

2.1.1. Scope of Function

The overall functionality of a program (application) is defined by the totality of the various features that are available, in principle, at the user interface and that are referred to in what follows as functions or commands. It consists of all the features implemented by the manufacturer, possibly supplemented by other functions defined by the user. The overall functionality need not always be directly accessible to the user, which is why our definition contains the phrase "available in principle". The set of functions to which the user has direct access is the current functionality in a given situation. The current functionality is a modifiable[1] subset of the overall functionality. Direct access means that the function involved can be called directly by using (one of) the customary activation technique(s) provided for that particular application. In the menu- and icon-based systems that now predominate in the PC world, this means that current functionality refers to the commands and functions callable by making one menu selection (not necessarily at the top level), or striking one key (combination), or clicking one icon. It may seem odd that this definition lays stress on a single step. This is because current functionality must be modifiable and therefore must contain the functions for changing its scope. If the definition included multi-step sequences of actions, such a sequence could itself modify current functionality to include any command of the overall functionality, thus making any distinction between current and overall functionality meaningless.

Enabling users to modify current functionality is an important step toward making systems individualizable. However, with possible the exception of adding macros, the current functionality of most systems is fixed, and therefore identical to the overall functionality. A very crude and insufficient form of adaptation consists of switching from a full to a short menu set. The scope of functionality in each set has already been defined by the manufacturer. This option can be found, for example, in earlier versions of WORD and EXCEL, for both DOS and Macintosh. Any commands not contained in the short menu set are also not accessible by the key shortcuts assigned to them.[2]

Only a few systems offer options for redefining their current functionality in such a flexible way that a task- and user-specific adaptation can be achieved.

---

[1]   If current functionality were invariable there would be no point in making a distinction between current and overall functionality.

[2]   The main object of having such rough settings only is obviously to avoid confronting the beginner with the full range of total functionality, which would make orientation difficult and might lead to a feeling of insecurity.

Examples are the WORD versions for Macintosh (upward of Version 4) and for WINDOWS (referred to in what follows as WINWORD). This corresponds to the fact that the initial current functionality of these two programs, as they are configured by the manufacturer, by no means comprises all the functions of their overall functionality. The user can introduce each function of the overall functionality, in principle, anywhere in a menu according to his preferences and allocate an optional key shortcut to it. In addition, a (key) shortcut may also be agreed to for functions that have no menu entry (i.e. shortcuts and menu entries are independent). The converse is also possible: Commands may be removed from menus or lose their shortcuts. The procedure for customizing functionality does not differ from normal system use. The user calls a special menu command[3] that brings up a dialogue box where he can make the appropriate settings (i.e., he employs a technique familiar to Macintosh and WINDOWS users).

This aspect of *method* distinguishes these two programs from the spreadsheet programs EXCEL (versions for Macintosh and WINDOWS), QUATTRO PRO, and WINGZ. In these applications, too, it is possible to customize current functionality, but the changes must be programmed in the application's macro language. In WINGZ, the idea that current functionality is generally a subset of total functionality must be qualified: This is only the case where the program is regarded as menuoriented. Besides function access via menu and key combinations, WINGZ has an always available command-language interface enabling the user to directly access all functions implemented by the manufacturer. So, for users able to employ that access form, there are no restrictions to current functionality.

So far, we have considered application-wide adaptations. A further step would be to make current functionality document-specific. Then, one application could have different adaptations for different tasks. Features of this kind are offered by WINWORD. The adaptation and, in particular, the definition of current functionality, can here be made specific to different types of documents. Every type is represented by a special sample document called a *template*.

Templates are specimen documents that the user can create and edit in nearly the same way as normal documents. They may contain (formatted) text and expressions (called fields in WINWORD terminology) that are references not only to text portions at different places of the document but to all information available in a "WINDOWS world", including data and pictures from doc-

---

3   This command is similar to other menu commands in another important respect: The user is able to remove it from the menu. If no other accesss to adaptation is available, the user can (accidentally) lock himself in to a particular functionality, and must reinstall the system to re-establish this option.

uments in other (WINDOWS) applications. Such external references can be updated at any time, so that this is a dynamic link. "Normal" documents are instances of templates. They inherit text and expressions from their templates and evaluate the expressions in specific situations (e.g., for a printout or on command), thus replacing the expressions by the results of the evaluation. When any (normal) document is generated, the user is asked to quote the template for the new document.[4] The current functionality available for processing the new (or an opened) document, including its accessibility by menus (structure and option names) or by shortcuts, as well as other properties (e.g., available [print] styles) reflect the adaptations made for the template involved.

EXCEL upward of Version 3 also has templates as specimens for new documents. Assuming proper configuration, the EXCEL user, too, can select a template other than the default, on generating a new document and is then given a copy of this template as his current working document. Using a mechanism for automatically activating macros when documents are generated or opened (described in detail in Sections 2.1.3 and 3.4), and supported by the power of the EXCEL macro language, the user can set up a template-specific functionality here, as well. However, this will then apply to all currently opened documents, even if some of these are instances of other templates. Another difference vis-à-vis WINWORD lies in the method: As just mentioned, in EXCEL macros must be used, which means programming instead of direct interaction.

So far, we have considered functional scope *within* applications. In many cases, however, some functionality is available to the user independent of the program currently being used even in a single-tasking environment. The user can access this "external" functionality without closing his application. This functionality normally has to be installed in a prescribed way allowing the (end) user to determine its scope. As a result of the installation procedure, the external functions are available after each system start-up. So, selection and installation of external functions can be viewed as a contribution toward adapting the system to the user's tasks. The collection of *desk accessories* in Macintosh is a case in point. Its functionality can be configured by choosing and installing the appropriate components. Results obtained using this external functionality can be brought into any application by means of operating system features; for example, results of operations with a calculator tool can be inserted into a document of a text-processing program.

In the DOS area, a similar extension exists, namely, resident programs loaded automatically when booting (e.g., SIDEKICK Plus). However, these can only be called by key combination and offer a fixed range of functions.

---

4   Default is the normal template, which contains no text or expressions, but merely provides a minimum set of styles and defines a default for current functionality and its accessibility.

This possibility of leaving an application while retaining the work context in order to perform an interim task using a defined range of external functionality and possibly transferring the result to the main application, is of particular importance for single-tasking environments. In window-oriented multitasking systems, this is a standard feature, and the scope of (external) functionality is freely definable by the user (i.e., he procures the required programs). So what can we still regard as configurable (in the longer term) under these circumstances if, in principle, any desired features can be used in an ad hoc fashion? With regard to the scope of the functionality, hardly anything, but we could imagine options for setting up very easy access to frequently used features, with the need for some of the features being possibly dependent on context, as well. If we note, for example, that a given task context requires frequent access from the main application to certain other applications, we might require that the other applications, too, be automatically loaded when the main application is loaded. Certain preconditions for this are offered in applications whose macro language has commands for starting other applications (or for setting up communication channels to them) and that also permit automatic execution of a macro (specifiable by the user) when they are started (see also Section 3.4). These preconditions exist in WINDOWS applications, but also in Macintosh EXCEL, for example. However, these configuration options, which primarily concern accessibility, should really be discussed under the heading "Interface".

User-Defined Commands

Recent versions of the applications being discussed here—with a few exceptions (e.g., WORD 5 for Macintosh)—allow sequences of user actions, including function calls, to be compiled under one name and made known to the system as a new function, which can then be called in one step. Such sequences are usually referred to as *macros*. Macros that are directly accessible to the user are part of the current functionality.

Where macros are derived from functions that are available at the user interface, they do not involve any fundamentally new functionality; by definition, anything that can be done by macros can also be done using a sequence of single calls for the various functions. This is also true, in theory, if the macro language additionally contains control flow statements, because the effects of decision making and repetitions in a macro can, in principle, be achieved by appropriate user behavior. In practice, however, the limits of such a user-charged simulation will soon be reached. So macros, under the proviso just mentioned, modify current functionality merely by forming larger units, but do not extend overall functionality. This is no longer true, however, if macro language power goes beyond a function range that is interactively accessible or if it provides an interface for connecting external routines to implement otherwise unobtainable features. Such features can then only be used by calling a macro. The possibility of adding external routines is available (e.g., in the two EXCEL variants, Macintosh and WINDOWS versions). Apart from this latter case of a genuine functional extension, the adaptation potential of macros lies mainly in their ability to step-up task-related efficiency by providing an appropriate current functionality whenever the same command sequences recur in the course of performing similar tasks.

From the user's point of view, macros should behave like ordinary functions, hiding their internal structure when they are used. In particular, the following properties are desirable:

- Macro functions should be called like any other functions including callability by, menu.

- Each macro should behave as a unit. This means specifically that they must offer

  - resettability as a whole, possibly also to intermediate states for provision of parameters (Undo and Redo support);

  - presenting of dialogue boxes and error messages dedicated and tailored to the macro as a whole rather than to he dialogue boxes or error messages of component functions; on the contrary, these must be suppressible.

- It must be possible to bring information about macro functions into the application's own help system. On-line help regarding macro functions is almost more important than help on original functions, because the former are not described in the system literature, so that their purpose and their call conditions may be forgotten easily.

At present, we know of no system that meets all these requirements. However, all the objectives, with the exception of Undo, are available in the EXCEL variants. Special mention should be made in this connection of the NISUS ability to integrate macros into the Undo concept (i.e., to reset even to a state prior to macro execution).

Add-Ins

Occasionally, the set of functions in an application can be extended by integrating external components. In the simplest case, these are macro files provided by the manufacturer or by another supplier for copying into certain directories. The loading procedures for the application (see Section 3.4) integrate the macro functionality into the system. Complete integration for the user exists where, as in EXCEL, these macros can be called via menu entries generated for this purpose, so that the user can no longer distinguish between a macro and a "genuine" command. The integration is less complete if an existing macro list is simply extended.

In other cases, the external components are separate programs with appropriate designs and filing conventions to ensure that their functionality is also accessible from other programs. In WINWORD 2, for example, such an extra program—say, a graphics application—can be selected and called as an option of a menu. In AMI PRO, a separate Mail menu is included in the menu bar when CC:MAIL, a product of the same manufacturer, is installed.

2.1.2. Defaults for Function Execution and Object Attributes

The impact of commands is determined in many cases by parameters that need not be—or are even incapable of being—explicitly set when the command is called. In executing the command, the program falls back on defaults. In many cases these defaults are changeable and therefore constitute an important aspect of adaptability.

Using defaults on command execution is a familiar principle. One aim is to avoid any overloading of a command with parameters when it is called. We need only think of the data input needed for a document printout (e.g., margins, headers and footers, pagination). A command with such a large number of mandatory parameters would be unmanageable. Above all, however, most of these parameters would have the same values in the majority of cases, suggesting that assignments are relatively independent of individual calls. This being so, the obvious step is to make the assignment independent of the command call. There are also cases where the same parameter values are used on different occasions. A word-processing program, for example, is only able to display a document in WYSIWYG (What You See Is What You Get) layout on the screen if it knows the values of all parameters that control the line and page makeup for printing.

Resetting parameters that are used as defaults is by no means always an adaptation. The situation can be analyzed in a little more detail taking as an example the printout of a document by WORD (Macintosh version). When the Print command is called, a dialogue box appears with a series of parameters. These parameters behave in different ways, however. The parameter (number of) *Copies* always has the value 1, regardless of the selection made by the user at the last activation. By contrast, the parameter *Print hidden text* does retain its modified value. So only this second parameter is adaptable, and the user can set a new default with each change. In this case, there is no possibility of differentiating between situational overwriting, while retaining the default value, and a new setting of the default value. With the *Copies* parameter, only situational overwriting is possible, and there is no option allowing the user to alter the default value itself. The criterion for a parameter setting to be regarded as an adaptation, therefore, is that it can be made *call outlasting*. Resetting layout parameters of a document provides another example: Where the new parameter values affect only the current document, we cannot speak of an adaptation; it would only be an adaptation if the changes also affected the layout of newly generated documents.

We found no consistent strategy determining whether an assignment is to be regarded as setting a new default or merely as situational overwriting. Even for variants of the same kind of application from a single supplier, there are inconsistencies. For instance, the already mentioned *Copies* parameter can be assigned a default value by the user in the DOS version of WORD 5.5, but not in the Macintosh version. The same is true of the Print selection.

The lifetime of defaults may vary. The most common case is "lifetime pending revocation". These long-term defaults may be overwritten by shorter lived defaults. A customary form of specialization allows parameter values other than the standard defaults to be set for (the remainder of) a session without forgetting the global values. Such parameters exist in NISUS and QUATTRO PRO.

In the following, we take a look at some examples of adaptation concerning default (re)setting that are characteristic of the types of applications considered here. One instance in the word-processing field is the possibility of activating and deactivating the paragraph control in the page makeup or the background page makeup of a document. An option that may be not immediately recognizable as a default setting is the creation of user-specific dictionaries. Nonetheless, this is a case of user-oriented influence on the *Speller* function in a very specific, almost evolutionary, way. In database systems, too, there are a number of varietal defaults such as:

- Setting limits on admissible field values for amendments to existing, or the input of new, records
- Rules (i.e., automatisms) for assignments to fields, to which no values were explicitly entered during data input
- Provision of a prior "filter" in output or display and/or further selection of records
- Definition of the (sub)set of fields to be displayed and their order.

These adaptation options exist (e.g., in the ADIMENS GT plus database system). One characteristic feature of spreadsheet programs is that they usually have options for setting the accuracy of calculations and defining the conditions for triggering a recalculation of the spreadsheet (automatically, automatically under certain conditions, or manually).

Adaptation options also exist, of course, in the case of application-*non*specific functions. Thus, practically all applications allow the user some layout control during start-up; often, the ubiquitous Save command can be conditioned to save the old version of the document as a backup copy in each case of (parameterless) activation, with any existing copy being overwritten by the new version.

Defaults are especially effective in the creation of new objects. Only in the rarest of cases will all the properties of the generated object be explicitly specified at generation time; instead, the system has recourse to defaults. Often, an object can be created in several ways[5], and there may not even be an actual generation command (e.g., in the creation of a new paragraph in word processors, where the user just hits the Return key). This being so, it may be more reasonable to view the defaults becoming effective in such cases as being attached not to the act of generation but to the generated object or, more precisely, to the object type. So, where options exist for changing these defaults, they primarily concern the object type and not a command. Any such adaptations become apparent and, hence, externally effective only when a new object is generated, however this is done. All the same, this does involve modifying the effect produced by certain commands or actions, so that we can speak in such cases, too, of adapting functionality.

The adaptation options for the properties of newly generated objects may be illustrated by using word processing as an example. In nearly all variations of this application type, the (normal) paragraph is an entity to which attributes can be attached. At the moment of generation, the still empty paragraph is automatically assigned a number of properties (letter style, letter size, display, line spacing, indentation, etc.) that affect the appearance of the paragraph text in the printout (and on the screen, depending on the degree of support of the WYSIWYG principle). As a rule, the program borrows the relevant properties from the preceding paragraph. For the first paragraph of a new document, however, there is no preceding paragraph, so that more global defaults come into play. Under the hereditary rule, they also determine the appearance of subsequent paragraphs until explicit formatting is undertaken. The set of parameter values used automatically by the system in such a case is frequently called an *automatic style* (e.g., in the WORD manuals). We also use this term. Moreover, it is essential for an automatic style that the values of the parameters involved can be altered by the user. Regarding the automatic use of the style, an alteration means a redefinition of defaults. The idea of automatic

---

5   For example, a title in WORD can be formed either by entering its text as a normal paragraph and assigning the appropriate style afterward or by entering the text directly in the outline mode.

styles for paragraphs is found (e.g., in AMI PRO, NISUS, and WORD), although implementation varies.

A plain text paragraph is not the only kind of paragraph to which automatic styles apply; other object types for which separate automatic styles exist in some systems include headline (at a certain level), header/footer, footnote text, footnote sign, and so on. The example of the footnote sign shows that automatic styles may also exist for pure character objects, where they have a specific status known to the system. The upshot of all this is that where systems have a wide range of object types with automatic styles assigned to them (AMI PRO, WORD), it is possible to design a very individually structured document without explicit formatting.

Although further moves toward autonomy in parameter settings may offer convenience, reducing (possibly to zero) the effort needed to feed in parameter values when commands are called, there is also a risk of certain effects being produced that the user doesn't want, because he is unaware of current defaults when starting work and because they were not displayed for his information. For example, the parameter *Print hidden text* in WORD 5.5 (DOS) is not visible in the dialogue box shown when the Print command is called, so the user must take the initiative and ask for a display by taking explicit action. Yet this parameter is a default (i.e., its effect outlives the call). If the user has printed out some documents in this state that have no hidden text, he may no longer be aware of the present setting. Later, when a document containing hidden text is printed out, the hard copy may not be what the user expected.

2.1.3. Delegation Options: Triggers

Some systems offer at least rudimentary options for delegating the execution of certain functions to the system in order to obtain automatic execution, without explicit user action, as soon as certain conditions are met or specific events occur. In the general case, the user specifies the actuating condition or triggering event and defines the response required. Such condition and response pairing, here called *triggers,* provides a considerable adaptation potential, because it can give a system a very specific dynamism of its own. The exaggerated use of triggers, however—for instance, such that they are mutually dependent—can lead to unforeseeable effects and unpredictable system behavior.

Triggers of a such general nature, for which the user can freely define and combine both the triggering event and the response, are not, however, a very common feature of the applications considered here. Apart from the cases described in the next paragraph, most applications tend to offer triggers as fixed pairs, with, at most, the possibility of setting parameters. A case in point is seen in the options offered by some word processing programs concerning automatic saving of edited text. The user can make one of the following choices: No automatic saving (trigger is switched off), automatic generation of a dialogue box asking whether the document is to be saved, or automatic saving without asking. If the trigger is active, he can specify additional details of the triggering event, such as time passed (DOS-WORD, WINWORD, WORD-PERFECT) or the number of key strokes (NISUS) since last save. In the database system ADIMENS GT plus, opening a database can be coupled with the loading of further files defined by the user to activate various previously set standards. Another example of a ready-made trigger provided by the system, where the user is only left to decide whether to deploy it or not, is the option of automatic calculation often found in spreadsheet programs. The triggering event in this case is a change in those cells whose content is needed to calculate formulae (in other cells). Once the changing action is completed, the table is automatically recalculated if the trigger is in force.

In more flexible cases, a (user) macro can be taken as the response part of a trigger, which is then often called an *autoexec* macro. This option is offered by almost all applications under discussion here, wherever the formation of macros is supported at all. Differences exist in the nature and variety of the triggering events. High specificity is offered (e.g., by the option of linking macros to the opening and closing of documents in EXCEL, NISUS, and WINGZ); that is, triggers can be made document specific. Specificity is more limited in WINWORD and (DOS) WORD; in the former case, the autoexec macros have a template-specific definition, namely, for the events "Create document", "Load document" and "Close document". In WORD, they are linked to the opening of macro or module documents.

In some systems, the loading of the application itself can be used to trigger a macro (NISUS, WINGZ, and WINWORD and DOS-WORD). This provides the user with a powerful adaptation mechanism, in particular, when the macro can generate a working environment with a task-specific functionality and interface immediately after application start-up. In the case of EXCEL, the same effect is produced by a combination of two triggers: the first one opening the documents (including macro sheets) of a specific directory on application loading, and the second one being an autoexec macro, which is attached to one of the automatically opened files. A similar option is offered by QUATTRO PRO, but with the reservation that only one macro and only one start-up file can be defined. Further remarks on this subject (loading an application as triggering event) can be found in Section 3.4.

The two EXCEL variants (and WINGZ, too, to some extent) provide a few more widely usable events besides those already mentioned. For example, macros can be linked to preset times, to the striking of certain keys or to a change in the active window, to name the three main such events.

Triggers are automatisms set up—or at least put in force—by the user himself. Mention was made earlier of the autonomous operation of automatic styles in the creation of new objects, especially paragraphs, as a special form of setting defaults in some word-processing programs. It might be asked whether this could not be regarded as a trigger. In fact, what is specifically missing here is the previously mentioned facultative element, because the existence of an appropriate automatic style is indispensable for the generation of an object.

2.2. User Interface Adaptation

So far, we have been considering options provided for modifying functionality. In the following section, we concern ourselves with the possibility of adapting the properties of the interface. In view of an interface's intermediary role between user and functionality, the properties are: accessibility to the functions of the system, the dynamics of the interface (timing and selection of messages to be displayed, including error messages), and screen layout.

2.2.1. Accessibility to Functionality

The standard way to access functionality in today's PC software is by menu. Often, the chief functions in an application are also given key combinations. Another recently established accessing technique in the latest versions of various applications involves a symbol or button bar. This is a screen area (sometimes in the form of a separate movable window) containing click-sensitive symbols (icons), each representing a command. A symbol bar can be regarded as a permanently open menu with icons as options instead of words. The icons are designed in such a way that the command can be intuitively inferred or is at least easily recalled by the user. So, here we have an obvious potential for adapting accessibility by amending the menu structure, renaming menu options, modifying and reassigning key combinations, and restructuring the button bars.

We start by considering *application-internal* adaptation features relying on the *normal* accessing techniques implemented in the interface of a particular application. We consider adaptations that can only be realized by *macro programming* (i.e., when simple macro recording does not suffice) to be a special case, although the options opened up in this way are often quite far-reaching. In the case of EXCEL, QUATTRO PRO, and WINGZ, for example, these options allow the user to completely rename and refashion the menu structure, possibly including assigned key shortcuts. The question is, however, whether, in view of the previous knowledge this method requires, this technique is really feasible for a wide range of users. Therefore, we do not consider this case further in the rest of this section. The corresponding methods are dealt with in the section on methods.

Menus

With the exception of the special feature for integrating macros, more general options for restructuring menus were found among the applications considered here only in Macintosh WORD since Version 4 and in WINWORD. Because they are quite extensive we will describe them in some detail on the basis of the features found in WORD. To begin with, we have the possibility of forming new commands that are special cases of general ones. Specialization is achieved by attaching fixed values to parameters that otherwise must be assigned interactively, thus avoiding a dialogue box popping up when the function is called. The resulting (now parameterless) commands are incorporated into the menu of the initial command. Exceptions are Open, Glossary and Styles, because new commands derived from these commands (e.g., for opening a specific document) are incorporated into a new menu (Work). Using commands created in this way bypasses the intermediate step of calling a dialogue box. When we consider, that, for example, nearly all the options in the window of the Format/Character command can become separate commands (in the Format menu), the drawback to this procedure is also obvious: Menus can become too large and unmanageable. Such user-generated commands can be removed from the menu again, of course.

The most comprehensive options, however, are offered by the Commands command in the Edit menu. This allows the user complete freedom to allocate all commands implemented by the manufacturer to the menus. Where a command not yet contained in the menu structure is to be integrated, the system proposes a target menu, but the user is not bound by this and can specify another target. The name of the new menu option is identical with the name of the function in the inventory of commands; renaming is not possible. A command can also be removed from the menu, of course.

All the options listed in the last paragraph for Macintosh WORD are also offered by WINWORD. In addition, however, the names of the menu commands may be freely selected (i.e., the user is not bound by the terms employed in the basic inventory). WINWORD also allows the user's own macros to be integrated into menus. Because, as was described earlier, in WINWORD adaptations can also be made document-type specific (i.e., template-specific), this means that the same function can be given different names in different contexts (e.g., technical terminology).

Key Shortcuts

In menu-oriented systems, a *shortcut*, as the name implies, is generally regarded as a quick way to activate a menu command. Typical examples for handling key shortcuts this way are NISUS and QUATTRO PRO (DOS). Special commands exist for assigning (new) shortcuts to all menu commands. As a consequence, the commands that are accessible by shortcuts are always a subset of those callable by menu. Remarkably, it is possible in NISUS to agree on a string of up to three characters as a shortcut. This obviates the need to fall back on unsatisfactory mnemonic shortcuts where these may be made up of one character only.

The situation is different, however, in WORD for Macintosh and WINDOWS. Here, shortcuts are no longer bound to menu entries. Rather, they constitute an independent access form. The inventory of the functions to which shortcuts can be assigned is the entire set of functions of the application, and not just those already accessible by menu. So, in WORD, menu entry and key assignment constitute combinable yet independent ways of accessing current functionality, so that a function may be made accessible by key combination only. The situation is illustrated in Table 1.1, which shows part of the set of basic commands and the available accessibility forms, as they are in the original configuration of (Macintosh) WORD 4, making it clear that each of the two access forms and even their combination cover only part of the total set of functions. Some functions are accessible in both ways (e.g., Document), some via only one avenue, and others cannot be (directly) accessed at all (Delete Rows). As with menus, in WINWORD key shortcuts can also be made template-specific.

If applications provide an option for recording macros, as is the case in practically all spreadsheets and in most word-processing programs, and if shortcuts can be assigned to these macros, this provides another way to assign shortcuts to commands.

Symbol Bars

Symbol bars have several associated adaptation features. To begin with, they are frequently presented in windows of their own, so they can be placed and sized to the user's convenience. A more important aspect is likely to be their configurability, which can be more or less flexible. It always includes defining the set of commands to be made available in the symbol bar, but the basis from which such a selection can be made varies: In WINWORD 2 it is the overall functionality, whereas in Excel 4 it is restricted to a subset predefined by the manufacturer. Moreover, as in the case of EXCEL, the functions associated to the icons in the symbol bar are sometimes special cases of those callable by menu, in that they have a fixed value assigned to a parameter. Another aspect of adapting symbol bars is the relation between the icon and the function performed; this relation can be altered in most (configurable) symbol bars (i.e., a given icon can be assigned to a different function and vice versa). In most cases icons can also be assigned to user-defined macros. This seems to become users' favorite method for making macros directly accessible. A pioneer in this area is Lotus 1-2-3 (WINDOWS version or DOS version 3.1), where, in addition, the user can generate icons of his own.

Command Name	Key Shortcut		Menu
Delete Forward	Command Option	F	
Delete Forward		Del	
Delete Next Word	Command Option	G	
Delete Previous Word	Command Option	Delete	
Delete Rows			
Delete...			File
Demote Heading			
Document...	Command	F14	Format
Dotted Underline	Command Shift	\	
Dotted Underline	Option	F12	
Double Space	Command Shift	Y	

Table 1.1:   Call options for various commands in the basic inventory of Mac WORD 4.

In recent versions of some applications, a series of symbol bars are available (EXCEL Version 4, and QUATTRO PRO WINDOWS version), each of them being configurable as already described. The user can decide which of them should be visible in a given situation, thus enabling him to confine the display to those he needs for his present work situation.

External Means of Interface Adaptation

In addition to the just-mentioned internal adaptation options, we also have *application-external* mechanisms for adapting application interfaces. The first to be discussed here concerns external *macro tools* (not to be confused with application-internal macro features). These tools are supplied either as part of the basic software (MACROMAKER[6] as a Macintosh utility and RECORDER as a WINDOWS component) or as separate programs (QUICKEYS and TEMPO, both for Macintosh). Their operation is described in more detail in Section 3.3. The point that interests us here is that such tools can (also) be used to create macros that merely execute a menu command (possibly with a particular parameter assignment) and that can be called from the keyboard. So, they offer an always available option for defining key shortcuts. This is done without modifying the code of the application itself. For each application, a separate set of such shortcuts can generally be set up, so that it is possible, in particular, to make typical standard functions (Open, Close, Exit, Cut, Copy, Insert) always accessible via the same user-defined key combinations, even if the commands involved have different names or shortcuts in the different applications.

For program development in Macintosh and under WINDOWS, it is possible to construct programs that have a special modular character, so that they can subsequently be modified to a certain extent using special tools. The portions open to modification are data structures called *resources*. Accordingly, the modifying tools are called *resource editors*. This is of importance for the present discussion because menus (and dialogue boxes) of applications, too, can (but need not) be implemented for the platforms named in the form of resources. Thus, program developers can in effect provide adaptation options without explicitly writing them. Using the editors named (e.g., RESEDIT for Macintosh or RESOURCE-WORKSHOP for WINDOWS), users can rename and remove menu commands and overwrite, reassign, or withdraw shortcuts. Use of a modern resource editor makes no special demands on user skills; on the other hand, this is not without its dangers, because some modifications only affect the surface and not the lower levels of an application: the result may be some completely unexpected system behavior. Use of tools of this kind does not appear to be very widespread among end-users, possibly for this reason. Application examples for RESEDIT are found in Section 3.2.

Special Provisions for File Access

One special case of an interface adaptation is modifying the accessibility of program files for start-up (or of data files for opening, with simultaneous start-up of the application concerned). Starting a program is inconvenient if its file is hidden somewhere in the hierarchy of the file system, because it requires "climbing through" the directories to the location of the desired application file. There are several techniques for simplifying this procedure without shifting or copying the various files. One technique long familiar in operating sys-

---

6   MACROMAKER was not yet available for Version 7 of the Macintosh operating system at the time of this writing.

tems is that of *linking* or setting up *alias* entries. The user may create several entries (icons) for a file object, which may be—and usefully are—filed in different directories; they can then have the same names as the original. The generation of an alias is not duplication of a file: Each alias always refers to the original file. In handling, there is no difference between an alias and an original entry; in particular, a file can be opened by a double click on one of its alias entries. Macintosh, for example, has had alias generation since operating system Version 7.

An alternative to alias definition is the user providing his own independent but parallel access structure—for example, in menu form—the sole purpose being to ensure access as directly as possible to certain files that he has chosen, regardless of their position in the directory hierarchy. In WINDOWS, such an access structure (although only for program files) is an integral part of the system (Program Manager). In addition, system extensions are available that, in theory, allow the entire inventory of files to be rearranged for access purposes (e.g., APOLLO for Macintosh).

Access to files can be simplified, of course, using the macro tools mentioned earlier (with reservations in the case of MACROMAKER), by recording and creating a macro that consists of an Open command for a particular file. Calling the macro (possibly by key shortcut) opens the file concerned regardless of its position in the file hierarchy, provided that its position has not changed since recording the macro.

Configuration Files for Program Execution

When calling a program by direct manipulation, such as via icons, there is no way to pass parameters. If, for example, a program is called to access remote computers, a link must first be established with that computer, with the user providing the appropriate (symbolic) address. In addition, specific parameters may also have to be set for the link to be built. Some programs permit such work configurations to be saved as separate files. The benefit for the user is not only that he can load such a configuration file again in a subsequent session, but, most of all, that such a file can be used to start up the application itself, with the saved work environment being automatically restored by the start-up procedure. These options are offered (e.g., by the program TELNET for Macintosh). Another example is NEWSWATCHER (also for Macintosh), which reads and posts messages, with a user-specific working environment being defined by selecting relevant groups. When the program is called by way of such a configuration file, the user is (initially) offered only those selected groups, the display indicating if any group has messages he has not yet read.

2.2.2. Dialogue behavior

Adaptation options in interactive mode are rare. We have no knowledge of any system that, for example, allows the user to decide about the dynamics (when and which details) of Acknowledge or Help messages or about the extent to which defaults are used in dialogue boxes, so that their display might be suppressed. However, mention should be made of the option of switching into a "help on" mode encountered in some systems. In the Macintosh operating system, Version 7, this feature is called *balloon help*. If it is implemented in an application (for the time being, this applies only to a few recent versions of programs), every object and menu item, when touched by the cursor in this mode, displays a speech bubble with a short explanation. In a variation of this feature, as implemented in ADIMENS (in the EXEC module), calling a command in the help mode first invokes a dialogue box with explanatory text on the command's effect; only then can a decision be made by the user on whether to execute or abort.

2.2.3. Layout

Options for adapting layouts are found in nearly every application. Users are usually much more familiar with their existence than with other adaptation options. An attempt to produce a complete list is not very helpful, so we confine ourselves to a few general aspects.

One important function of layout changes is to adapt a layout to individual aesthetic and design preferences. Color options are often available (e.g., to distinguish between various functional areas in a window). It is sometimes possible to alter the frame of a window (no frame, frame consisting of one or two lines). Rarer options include those for redesigning the look of forms, including dialogue boxes, wherever these are provided by the system and are not generated by the user. EXCEL systems for working with a database set up on a spreadsheet allow the replacement of the default-based data entry form by a user-created dialogue box. The latter can also be produced by direct manipulation, using an extra program (*Dialog Editor*).

This special EXCEL entry form is the only case we are aware of that offers options for modifying dialogue boxes (structure and layout) in an application using the program's own methods (even if only by using a utility program supplied along with the main program). Although dialogue boxes in Macintosh and WINDOWS environments do perform a central role in user–computer communication, we have found, no other internal adaptation options. Such options (position and size of the box itself, location, size, and lettering of its elements), however, are offered by the resource editors already mentioned in the section on accessibility.

Among the familiar options for adapting layout are those for switching the display of status information and icon bars on and off, mostly at the cost—or in favor—of the work space. Such modifications naturally also have associated effects on the accessibility of system features, particularly in the case of icon bars. In fact, the effects may be so extensive that it is moot whether the adaptation is to layout or to functionality. Some examples may illustrate this. If you suppress the button bars, you lose the directly manipulative access to some commands; in EXCEL, the scroll bars of the spreadsheet windows can be masked, but at the expense of the scrolling option with the mouse. In WORD 5.5, like other display areas the menu bar can be suppressed so that an intermediate step—pressing the ALT key[7]—is necessary to display the menu temporarily, if a command is to be called by mouse.

One last area for adapting layout concerns the definability of default formats for displaying data types, such as date, time of day, currency, weights, and measures. Such options were rarely encountered within applications. These features exist more frequently as part of the cross-application adaptation options described in the following section, often in the form of country-specific settings.

2.2.4. Cross-Application Interface Adaptations

The interface adaptations discussed so far are application specific, but a number of interface features can also be defined independently of a particular application. These are described together here, even though they belong to different categories (in the classification scheme adopted so far).

From the outset, one characteristic of the directly manipulative Macintosh interface was the ability to define certain interface features externally to, but affecting, all applications and, hence, specific to the computer concerned. Much the same is true of WINDOWS environments. These adaptations encompass features at an instrumental level mainly involving keyboard and mouse (e.g., the relation between mouse and cursor movements, the time allowed for a double click, the interval before the response of the Key Repeat function, but also the nature and volume of acoustic signals such as the control panel or system control parameter).

---

[7]   This key must be pressed when the menu is visible, as well, if the cursor is to be positioned by keyboard from the work area to the menu bar (i.e., a command is to be called by keyboard only).

Adaptation options also exist at another level, for country-specific display conventions, for example, especially for dates and times of day. Another area is the color scheme for generic interface objects like windows, menus, and dialogue boxes. We made mention earlier of autonomously settable parameters for the commands of an application, and we might not be going too far if we were to speak here of autonomously settable parameters for applications as a whole. It is clear that the potential for cross-application adaptations is all the greater the more the applications are able to use generic objects and generic data types (time, date). This is very evident in the DOS area, where suitable objects and resulting adaptation options have only been provided to a noticeable extent in the course of the graphic extensions to the operating system (e.g., WINDOWS, GEM, etc.). Of course, they can only be used by applications designed to suit these extensions.

Finally, mention should be made of a Macintosh-specific feature. Unlike DOS environments, where, because of the hardware equipment of many users, the switch to graphic user interfaces was always bound up with an attempt to offer full program functionality via the keyboard as well, Macintosh had a division of functionality between mouse and keyboard from the very beginning. A typical mouse domain in this respect is the manipulation of windows. Enlarging, scrolling, and so on can, as a rule, only be executed by mouse. However, special software has been developed enabling access to such features by keyboard, as well, independently of a particular application. One such product is the already mentioned QUICKEYS, which allows freely selectable key combinations to be assigned to the addressed operations.

3. Methods of Adaptability

In the preceding section dealing with content, it was not possible, for reasons of comprehensibility, to entirely ignore method. All the same, separate treatment of this topic seems justified because identical or similar procedures can be employed to handle adaptability in different specific areas, so that methods can usefully be discussed independently of particular uses.

Applications can be adapted both by methods internal to the application and by external tools. These are the subjects of the first two subsections. As special cases we consider the defining of user-built commands and control options at start-up.

3.1. Application-Internal Methods

Almost every program offers adaptation options. In the simplest case, we have functions for switching back and forth between two states. Cases in point are the commands for switching on and off the display of special characters or of rulers in word-processing programs. Although such actions often have a call-outlasting effect, users may not consider them to be adaptations because they are frequently used in an ad hoc manner, so that referring to them as an "adaptation" may seem to be rather extreme. A user will have the feeling that a modification is an adaptation primarily when the effects are regarded as long-term in nature. This feeling obviously also arises if special commands are available dedicated for adaptation, bundling several adaptation options and having a matching name (e.g., "preferences"). As far as method is concerned, the use of such commands differs in no way from the use of other commands and thus requires no further explanation.

Sometimes, options for (re)setting local or document-specific parameters and for (re)defining defaults are intermixed in one command. An example is the Format/Document command in WORD (Macintosh), and more precisely, its dialogue box. Here, the user provides particulars affecting a document as a whole, such as margins, footnote management, and so on. In this situation, he can decide whether the current values of the dialogue box are only to apply to the active document (document-bound settings) or whether the new values are also to apply from now on as defaults for each new document generated (application-wide defaults). Should he want the latter effect, he has to click the "Set default" button. In WINWORD, however, "Set default" in this situation means the default with respect to the corresponding template only.

As was said earlier, defaults may differ in the duration of their validity, that is, only for the current session or beyond. As a consequence, there are normally two command types in such cases. Commands of the first type reset the parameter values and make them defaults for the time being (i.e., for the current session only). Only a command of the second type (usually, there is only one, e.g., in NISUS or QUATTRO PRO) makes changes that remain in force beyond the present session.

In Sections 2.1.1 and 2.2.1, specific reference was made to the ability to flexibly redefine the scope of directly accessible functionality and to modify accessibility found in Macintosh WORD and in WINWORD. The procedure in WORD (Version 4) is explained in some detail by means of Fig. 1.1. Calling the Commands option in the Edit menu causes the dialogue box shown in the illustration to be displayed. The panel on the left shows a subset of the entire basic inventory of commands. When a command is highlighted, the elements to the right indicate whether an access form—and which—is available for this command. In the situation shown in the illustration, no call option exists for the highlighted basic command Delete Rows. This is indicated first by the fact that, for menu accessibility, an "Add" button is shown and a suggestion is made for the adopting menu (Document). So, there is no menu accessibility as yet. Second, the list panel under "Keys:" is empty, indicating the absence of keyboard accessibility. The user can now set up both accesses. If he wishes to establish a menu access, the system suggestion for the menu to be chosen is not binding; he can open the list panel containing the entry "Document" in the illustration to show all feasible menus, one of which he can choose. Where the user makes a choice other than the system proposal, the new command is always added to the bottom of the selected menu. If the system proposal is accepted, the system will position the new command at a default position in the proposed menu ("Auto"), but the user can also have it positioned at the end ("Append"). Either way, the command is placed in the menu with the name indicated in the list. Where a command has been selected from the list of basic functions and is already contained in a menu, the content of the menu button changes from "Add" to "Remove". So it is not possible to include a command in more than one menu.

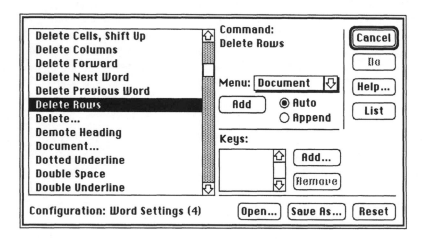

Figure 1.1:   Dialogue box for configuring the menus in WORD 4.

Where a (further) shortcut key is to be assigned, the "Add..." button in the key area of the dialogue box must be clicked; the system then requires input of a key combination. The new key then shows in the list panel (empty in our illustration). To cancel a key shortcut (whether system-given or user-defined), it must first be marked before the "Remove" button can be activated. It is worth recalling at this point that definition of a key combination does not presuppose the existence of a menu entry.

In WINWORD 2, rearranging menus and assigning key shortcuts are decoupled: the corresponding command forks into several branches, including, among others, one for adapting menus, one for assigning shortcuts, and, as a new feature, one for redefining the symbol bar. Each branch has a dialogue box of its own.

3.1.1. Definition and Invocation of Standby Parameter Sets

So far, we have assumed that specifying new parameter values or conditions is automatically associated with their coming into force. This does not apply in one important special case, described here as the definition of *constraint sets*. For certain purposes, several parameters often have to be reset together. Formatting a headline, for example, generally involves a whole series of parameters (character size, highlighting, upper margin, lower margin). To take this into account, some applications have functions that allow several parameters to be bundled and used as a package. The following aspects are characteristic of such a parameter package:

1. The various parameters in the package can be assigned values without taking immediate effect (specification of a standby assignment)
2. Such an allocation of values, as in 1., for the parameters of a package can be given a name and saved as a whole, possibly for use beyond the current session
3. An assignment saved according to 2. can come into force at once by a reference to its name.

A characteristic feature of the technique described here is the two-stage nature of the procedure: first, values are given as an alternative assignment, which then come into effect at a later time. Although the user can, as a rule, demand immediate effect when he inputs the new values, what matters here is that this is not mandatory. Naturally, there is no point in such a two-stage character in the case of a single parameter.

Options for forming such packages can be found, for example, in the more sophisticated word-processing systems, mainly for formatting paragraphs, but also (strings of) for characters or pages. Some spreadsheet programs (EXCEL, QUATTRO PRO, both upward of Version 3) offer similar features for formatting cell areas. A constraint set for paragraph parameters is usually called a *style* (WORD, WORDPERFECT, EXCEL) or *style sheet* (NISUS). Earlier (in Section 2.1.2), we spoke of *automatic* styles: These are constraint sets that the system assigns as defaults to certain types of objects. The techniques used to assign a style the status of an automatic style vary from application to application (where this option exists at all). In WORD, for example, it is solely the name that decides if a style is an automatic style or not; a certain name is required for each kind of text (plain text, headlines, footnotes, etc.). In NISUS, by contrast, the default style for all kinds of paragraphs is the style used in a certain document stored at a defined location, containing one empty paragraph only.

Another example is the option in the ADIMENS GT plus database system of defining complex standby conditions, combined by Boolean operations from single conditions on various fields, and making them effective as required. The entire condition then acts as a filter for input and output. In the same way, standby definition is also possible for various expressions evaluating certain fields of each record and thus defining virtual fields on demand.

One special and very far-reaching variation on the subject of constraint sets is that of sample documents. This is a special status that can be assigned to documents in some applications (AMI PRO, EXCEL, NISUS, WINWORD, Mac WORD Version 5). Earlier we called these sample documents *templates*. The crucial point about a template is the different semantics of the Load command: Instead of opening the document concerned, a copy is produced and then opened once a working title has been assigned. In WINWORD and in EXCEL, too, under certain conditions[8]—with the call of the New command, a dialogue box is displayed presenting the templates available, so that the user can refer to one of them in creating a document. This template technique gives the user broad options in preparing task-specific documents, with not only a predefined layout, but—and in particular—predefined content as well. This allows structured forms and blanks to be defined in word processing.

In most cases, it is possible to obtain the same effects using constraint sets by making manual alterations. Wherever this requires several functions, these can be combined into a macro. Occasionally, such macros are accorded the same status as styles on account of their apparently comparable potential. However, true equivalence only exists if the macros can be used not only to produce a change in the present, but also to make appropriate standard effects binding for the future, for it is the crucial feature of a constraint set that it applies in the future as well. Even trying to achieve the same effect by way of subsequent alteration may involve problems (e.g., when errors occur in data input only because the proper filter was not in operation). For later correction, the original data may no longer be available and may have to be re-input at corresponding expense.

---

[8]   The templates must be located in the EXCEL start-up directory .

3.2. Use of External Tools

In this section, we discuss how some of the external tools mentioned in the functional account (Sections 2.2.1 and 2.2.4) can be used. Our object is not to give detailed operating instructions, but to allow the reader to form an idea of the conditions governing sensible use of such tools. The best course is to explain the procedures using illustrations of specific interface layouts.

To start, we demonstrate two application options offered by the Macintosh RESEDIT program: menu modification and repositioning of a dialogue box. Figure 1.2 shows RESEDIT in a state in which the Options menu of the SUPERPAINT program can be modified. When a program file is opened with RESEDIT, the user sees the names of the resource[9] types (here an extract from Kasc to PiMi in the rear window, top left) used in the program. As a rule, there are several instances of each type, and they are identified by ID numbers. The various resource instances are accessed by opening their type (e.g., the Menu instances by opening the MENU type). The middle window now shows all the menus belonging to the application (again, only an extract is shown here). Finally, by opening the Options menu, the user is given the front window shown in the illustration, which now allows him to alter this menu. The structure of many resource types is binding for all Macintosh programs, and this includes the MENU type as well. This being so, RESEDIT is able to work on these resources and present them to the user in a familiar and easy-to-interpret form. The contents of the displayed fields and buttons can now be modified in the obvious way, so that the menu and its options can be renamed and/or key shortcuts can be reset or withdrawn. However, RESEDIT unfortunately also allows changes to be made to the interface that have implications for other layers of the application. Thus, menu commands can be canceled or reintegrated, not just inside this menu, but in others as well. Such an amended program is now practically useless, because the internal assignment of the called features is not performed by the command name presented, but by the position of the mouse click, although processing is on the basis of the original menus. In other words, the commands no longer have the expected features. If menus are abridged by removing commands, the result is that the last commands in such a menu can no longer be called because their original position can never be hit in the shortened menu; if this is done in the File menu, the program can no longer be finished, because Quit is normally the last command of this menu.

---

9   As has been described in Section 2.2.1, resources are portions of a program that can be altered even in off-the-shelf software.

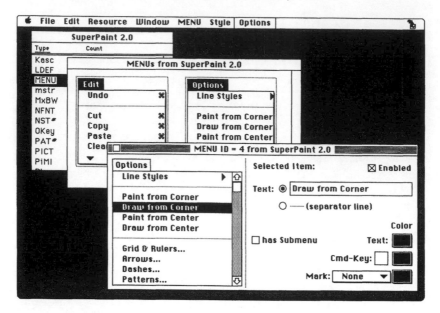

Figure 1.2:   Changing menu entries with RESEDIT.

Figure 1.3 shows the SUPERPAINT dialogue box with the content of resource 2000 of the DLOG type. The resources of this type mainly contain layout information for dialogue boxes as a whole. In this case, too, the program is able to interpret the information and offer it to the user in a form permitting change by direct manipulation. In the situation shown, the location of the box relative to the screen can be shifted by mouse. A subsequent call finds the dialogue box in the new position. The window type can also be defined (shiftable, scrollable, etc.).

The two examples show that the use of RESEDIT to make such modifications—which are definitely of interest as adaptations—need not place heavy demands on the user. However, if such a tool can also cause serious damage of the type previously described, it is not suitable for the end user.

Adaptations using resource editors like RESEDIT are made by modifying the program. Other external tools approach adaptations by providing a special operating environment; the program itself remains unchanged. This is particularly true of utility programs affecting, among other things, keyboard accessibility. These are upstream of the current application, as it were. They intercept the user's keyboard input, evaluate it according to previously established criteria, (e.g., replacing the internal representation of the input by another) and pass the result on to the application.

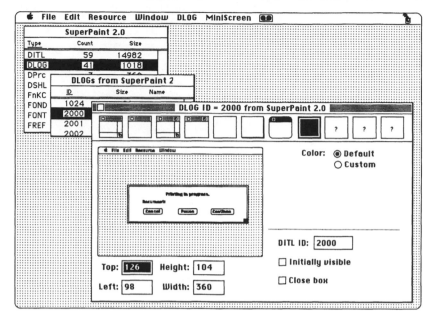

Figure 1.3:   **Changing the attributes of a dialogue box with RESEDIT.**

Examples of this tool type are RECORDER for WINDOWS or MACRO-MAKER, or TEMPO and QUICKEYS for Macintosh. The first three are straight macro tools, which are dealt with more generally in the next section. Although QUICKEYS also has features for creating macros, it offers much widerranging and dedicated options for agreeing on key shortcuts without having to manipulate a set of data (i.e.,macro recordings). Modifying keyboard accessibility using macro tools simply involves recording the striking of a key or a shortcut (if a command is to be executed) as a very short macro, which is then, in its turn, called by (another) key shortcut. All the products named allow shortcuts to be assigned to macros. RECORDER, MACROMAKER, and QUICKEYS even allow shortcuts without modification key(s). If the macro itself only involves striking a single (newly assigned) key, the result is a partial redefinition of the keyboard assignments.[10] Shortcuts agreed to using macro tools have priority over any others existing in the application.

Finally, some examples of shortcuts assigned in QUICKEYS without macro recording: Figs. 1.4 and 1.5 show stages in the definition of a key combina-

---

[10]   If the ANSI.SYS driver is loaded, the same effect can be obtained under DOS with certain ANSI escape character strings.

tion for calling a particular menu command in WORD. To do this, WORD
must be active. After installation, QUICKEYS is an option in the Apple menu
(and also a component in the *Control Panel* desk accessory), so it can be activated at any time. The QUICKEYS Define menu brings together all the options for assigning new key shortcuts. If this concerns a menu command, the
option Menu/DA is selected (Fig. 1.4). A dialogue box is now shown, calling on the user to choose a menu command in the application (Fig. 1.5). Then
the key shortcut itself has to be issued. Lastly Figure 1.6 shows the spectrum
of mouse actions to which key shortcuts can be allocated.

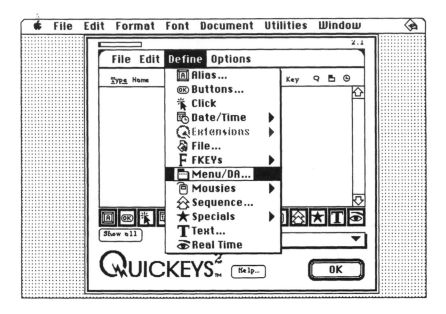

**Figure 1.4:   Defining a key combination for a menu
command with QUICKEYS - Phase 1.**

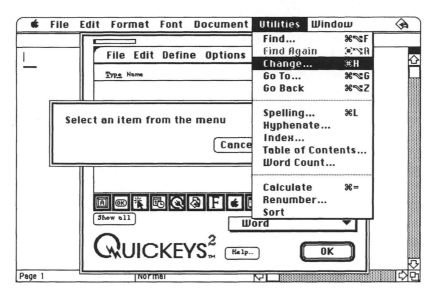

Figure 1.5:   Defining a key combination for a menu command with QUICKEYS - Phase 2.

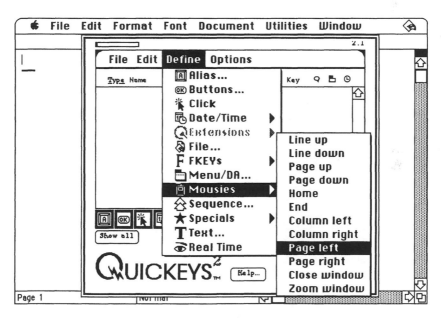

Figure 1.6:   Possible key equivalents for mouse actions in QUICKEYS.

3.3. Creation of User-Defined Commands

The functional aspects of user-defined commands were discussed in Section 2.1.1. Regarding terminology, we proceed as we did there and refer to them as *macros*. Feeling that the effect of a macro should be described as a unit, we have avoided the term in the heading, because this view may not be generally shared.

Macros can be created using both program-internal options and external tools, so that our remarks might be distributed between the two previous sections. For the sake of a continuous account, however, and in view of the importance of this subject, we have devoted a separate section to this topic.

In principle, there are two ways of obtaining macros; a third method is a combination of the other two. The first is to have the system *record* the user's actions. In most cases, the system provides two special functions for this purpose, namely, (freely named) Start Recording and End Recording. The first function switches to the recording mode, and the user's subsequent actions are recorded until he calls the function End Recording. The user must assign a name before or after recording; in most cases, a key shortcut can be agreed to, either additionally or alternatively. Some applications, however, always record the user's actions, so that there is no need for Start and End commands. The record is accessible to the user; in particular, copies can be made, so that if the record or part of the record is to be used as a macro, that part may be copied into a document provided for this purpose, and a name assigned to it to make it accessible for use. However, the user has to examine and analyze the record, in order to select the beginning and the end of the macro. This does presuppose a certain minimal knowledge of the notation used in recording, although intuition will reveal the meaning of many entries. All the same, the user is no longer entirely screened off from the (internal) representation of the macro. This way of creating macros has been implemented, for example, in LOTUS 1-2-3. It is also available in QUATTRO PRO, in addition to the first procedure described.

The recording method for creating macros is almost always available in a macro facility, although some restrictions may exist with regard to use of the mouse. Thus, mouse action may not be possible at all during the recording phase, especially in word-processing programs; instead, key combinations must be used wherever possible.

Use of the rigid command sequences that occur in a recorded macro is, in the nature of things, subject to narrow limits; the more specific the purpose of a macro, the more care must be taken in using it to ensure that comparable initial conditions exist (e.g., the same on-screen selections). This may involve problems in practical operations, so that such macros are mainly employed only for simple but widely used functions.

The second method of creating macros is (macro) programming. This means that the functionality of a system must also be accessible, at least for the purpose of macro production, via a command language, frequently called a *macro language*. If this language is complete, it must, in principle, cover the complete spectrum of features available in direct interaction. Wherever certain features can be obtained directly only by using the mouse, command-language equivalents must exist for these, as well. The required macros are then written by the user in this language.

Macro languages are available in almost all applications that support the creation of macros, but hardly ever in the application-external macro tools we discuss later. Macro languages usually contain not only the linguistic equivalents of the directly accessible functions, but other linguistic instruments as well. These primarily include constructs for defining control structures, for (interactive) data prompts, and for scanning the state of the active document. Even the formation of parameterizable subprograms is possible. So, macro languages offer options similar to conventional programming languages, but also have similar barriers to exhaustive use.

Systems with macro languages also usually have macro recording facilities, so the user is not forced to learn how to program with these systems either. In such cases, however, he has the great advantage that a macro produced by recording can be represented in the macro language, so that it is available for subsequent inspection and modification. The two methods can be combined, therefore. Users with some basic programming skills are able to extend the range of applications considerably by subsequently editingin case-specific differences (e.g., based on the conditions of the current environment) or by interactive calls for user input. It may be assumed that the existence of an initial basis in the form of the recorded macros considerably lowers the user's inhibitions with regard to programming, which in this case is of a supplemental character.

The macro languages found in some systems, such as EXCEL, QUATTRO PRO, WINGZ, or WINWORD, go beyond the options described up to now to include functions that open up far-reaching adaptation options: this is sketched here in the case of spreadsheets. Spreadsheets ought to be useful for many purposes involving calculations: They should be neither task-specific—financial accountancy, warehousing, etc.—nor sector-specific. In particular environments, however, specificity is often desirable, and this can be achieved—always with certain reservations, above all regarding performance—in the products considered here, namely, to transform the general tool to a task-adapted program. What we need for this purpose is, first, the ability to replace existing menus, either completely or partly, with user-designed menus.

This goes beyond mere restructuring or renaming; the (new) menu options can also be assigned user-defined functions in the form of macros (WINGZ uses the term *script* instead). In addition, it must also be possible to integrate user-defined and task-related dialogue boxes into the macros instead of the original dialogue boxes. Also, conditions must exist for macro-internal error processing to ensure that error messages from the various component commands are suppressed and task-related messages are output instead. When all these features are systematically used, the adapted application may be so changed in appearance that there is little to indicate its origins.

The two EXCEL variants go one step further. Here, the macro language offers the option of integrating externally produced binary routines into macros. These routines can be produced by general programming environments, so that system calls, too, are accessible and, therefore, any features that are required can be programmed.

In WINWORD, functional adaptation by macros can be made template-specific. These macros can be viewed as constituent parts of the template and are automatically available to all instances of a template (and only to them, if defined that way). So, it is possible, simply by activating (or creating) a document belonging to another template, to switch to a functionally different variant of the same program. WINWORD has global macros as well.

So far, we have been considering macros whose formation and use are part of the package in a particular application. For Macintosh, in particular, macros can also be formed using universal tools (or *macro tools*). *Universal* means that they can, in principle, be used along with any program (i.e., grafted on to it, as it were). The MACROMAKER utility is certainly the bestknown example being offered by the computer manufacturers themselves as an addition to the system software (operating system Version 6.0x). Other products are QUICKEYS (Version 2) and TEMPO, the latter offering the most comprehensive recording options. Following installation of the appropriate utilities, the macro functionality (recording, activation, etc.) is available either as part of an additional menu—both in the Finder and in active applications (MACROMAKER, TEMPO)—or as a new command in the Apple menu (QUICKEYS).

In the case of MACROMAKER and TEMPO, macros are always produced by recording. In TEMPO, commands from the TEMPO menu can be activated during recording to set up simple dialogue boxes displayed as data entry prompts when running, or to define rudimentary branching and looping constructs, the latter being oriented to the content of the clipboard as a comparative value (i.e., the clipboard can be used as a variable). The macro representation is not accessible to the user in either program. Subsequent editing, for which TEMPO offers certain options, is done within the scope of a special playback mode, in which what matters is timely interruption by the user to create the suitable initial situation for a change.

The real reason for the lack of a user-accessible representation of these macros is the level of their representation: Events are frequently recorded in a rather uninterpreted form; for example. a mouse click is basically represented as a pair of coordinates with a time marker. This method of describing user actions is also referred to as description on the *lexical* level and obviously offers no useful basis for subsequent editing by the user. Conversely, in the case of application-internal macros, the situation was fundamentally different; there, the recording was on the basis of commands, that is, on a *semantic* level.

A further consequence of recording on the basis of lexical events is that when directly manipulative input devices (mouse, graphic tray, etc.) are used, such macros are, to a high degree, dependent on the state of the screen. Only if the same screen conditions exist at both the time of recording and the time of use will the call produce the desired effect. This is particularly true of the activation of menu commands by mouse: The macro as a rule contains only the position of the mouse click. If the position of a command in a menu has been altered after, say, a switch from short to full form, the macro will activate the wrong commands. So, for application-independent macros that are based on lexical records of user actions, the advantage of universality brings with it a number of serious drawbacks.

One special case among the separate macro tools for Macintosh is that of QUICKEYS. Within the scope of its macro formation options, QUICKEYS allows the explicit compilation of sequences from previously agreed on and named actions[11]. This is done by selecting the required actions from the existing action sets presented in a special dialogue panel. Such a sequence can be altered again at any time; in particular, some commands may be specifically (re)configured. For example, the user can require that a button or window (dialogue box) be identified not by its sequence number, but by name. Furthermore, it is possible to introduce control structures that are oriented on the states of menus, buttons, or windows into a sequence by special extensions. Additionally, the current Version 2 also offers a recording option with two modes: as a sequence or as a real-time macro. In the first case, the result is available in the same form as it is in a manually composed sequence, and it has the same subsequent intervention options. QUICKEYS is therefore the only (application-external) macro tool considered here that gives the user an accessible representation of the recording. In the second mode, real-time macros, all actions, including otherwise ineffective mouse movements, are recorded in a purely lexical form that is inaccessible to the user, but reproducible by play-

---

[11]   These actions may be assigned shortcut keys.

back in real time[12]. Accordingly, our previous comments on the applicability of "lexical" macros also apply here. However, reservations must be made with regard to the accessible form as well where mouse clicks are used. The results may be unsatisfactory, especially when several options are clicked in the dialogue boxes. Another example is scrolling by mouse: the window for the display must be the same size as the window for the recording. By contrast, a menu command contained in a recorded sequence is unaffected by its position in the menu if, as in the default case, it is called by name.

The RECORDER macro function that has been added to WINDOWS since Version 3.0 avoids some of the serious drawbacks associated with purely lexical recording, while retaining general serviceability (within WINDOWS applications, at least). Although it also records primarily on the basis of lexical events, so that no editable representation of the macro is available, the record is enriched by identifiers of the application windows in which the manipulation takes place, including those that are created during the recording. As a default, this information is then used in the playback to check whether the same windows are open in analogous situations as were during the recording. In the case of nonagreement, execution of the macro is broken off. Together with the possibility of implementing recording and playback relative to the application windows (instead of the screen), this definitely reduces the risk of triggering wrong effects by mouse clicks even in a changed screen layout. This is due, in particular, to the fact that the menus of an application, including the current application, are not presented by windows at fixed screen positions, but in the current application window, so that within the same application and the same large window and window-relative recording or reproduction, menu commands can always be correctly selected when triggered by mouse, as well.

3.4. Conditioning at Start-Up

Many programs are designed to evaluate at start-up, a certain file, certain files, or all files in certain directories in a defined manner specific to that particular file. The start-up files contain information of various kinds, roughly classifiable as settings with regard to presentation, impact, and initial context for the program concerned (i.e., parameters for initialization); and as information about the execution of certain initial actions. Some examples are given in the following paragraphs. The usual precondition for a program to recognize its start-up files or directories is that these be available in a specific directory under a given name (i.e., the program examines predefined, normally invariable paths). This applies to both system and application programs.

---

[12]   Similarly to the TEMPO option.

The processing of start-up files has a long tradition in the system software area, especially in operating systems and in programming environments, and is used widely there. But even where the end-user is left on his own with his computer, this procedure is used both in application and in system software. The CONFIG.SYS and AUTOEXEC.BAT files, for example, are start-up files in DOS. The former mainly concerns adaptations to the available hardware and optimizes the system on a very technical level (e.g., number and size of buffers, drivers for physical or logical units), whereas AUTOEXEC.BAT is primarily concerned with aspects that take account of the user's needs. Besides defining a set of search paths for applications, the user can, for example, specify whether an application is to be opened automatically on booting. Most DOS applications that place certain demands on the system configuration modify these files in the course of installation, so that changes are made without any specific action on the part of the user. To a certain extent, this screens the user off from the technology involved. When the system is rebooted, it has been adapted to the needs of a particular program. All the same, conflicts cannot be ruled out between the requirements of different applications, especially when they are all resident.

WINDOWS makes very intensive use of start-up files. A large number of parameters affecting system configuration and presentation under WINDOWS are allocated values from the SYSTEM.INI, WIN.INI, and other *.INI files in booting. As in DOS, WINDOWS offers the option of using a start-up file (WIN.INI) to define which applications must be loaded automatically when booting. The user may also specify which documents should be opened within these applications.

The adaptation options using start-up files are also versatile in Macintosh. From the outset (i.e., at a time when only one regular program could be opened), the use of such files was intended to provide—in addition to the functionality of the current application—permanent access to a number of (utility) programs (calculator, clock, control panel—a tool for setting cross-application parameters, etc.) capable of being called at any time. This framework or external functionality is present under the always available Apple menu. This concept of the system also includes that this framework functionality be extendable, namely, it allows the user to add new features. This should be done (as a rule) not by user programming, but by integrating modules supplied by the manufacturer or by other suppliers as separate files. These so-called *Init* or *cdev* files (the names stem from inner designations of these files) need only be placed in a certain directory, the *System Folder*. (In Version 7, greater differentiation is necessary (i.e., certain subdirectories of the System Folder must be used.) So, this is a procedure familiar to the end user as well. The technical precondition for programming such integratable modules is the existence and accessibility of appropriate interfaces. The new features provided by such start-up files are mostly accessible via entries in the correspondingly enlarged Apple menu or as part of an extended control panel functionality. Some of these additional programs, especially macro tools, generate instead an additional menu in the menu bar of each application, including the Finder. A further accessibility technique is the generation of mouse-sensitive areas on the screen, containing "invisible" objects, mostly in the top left or right-hand corner. A mouse click in such an area makes the object visible and, thus, accessible. APOLLO,[13] for example, can be called in this way. In other cases, no direct accessibility exists at all for the new functionality, because the latter is used as a system utility function and, for example, provides the driver for a laser printer.

Start-up files with parameters for applications are frequently created when the application is installed or first started up and given default values. Usually, the settings that a user can make in his normal use of the system (e.g., by suitable menu commands) are stored in these files, for example, by overwriting the previous (default) values. Sometimes, such implicit action via the application concerned is the only way to alter start-up files. This is true above all for Macintosh applications (e.g., the start-up files of WORD and EXCEL). DOS and WINDOWS, however, often offer direct access to a start-up file in addition to the route via the application, by way of a plain ASCII file that can be manipulated by appropriate editors. Sometimes, this is the only way to get certain standards into a start-up file or to implement them there. Thus, changes to some settings stored in the WINDOWS file WIN.INI (e.g., movement tolerance in the double click, the colors of certain windows) can only be made by editor.

There are even cases where all manipulation of start-up files, including their creation, has to be made by external programs (editors); the application concerned only reads the optional file. In the ADIMENS database system, for example, this is the only way—again, adhering to the appropriate storage and naming conventions—to assign a filter or a computing rule as a default to each database. In the nature of things, adapting a start-up file using an external program has its drawbacks: The user has to leave the current application to make the change, and, as a rule, the changes made only become effective when the application is re-started, so that the work context is lost. Apart from the need to have command of another tool (the editor), which may not otherwise be needed for the current task, it is necessary to know the structure of the start-up file (i.e., the nature and meaning of the entries prepared or created). Manufacturers, too, obviously assume that most users will not—or even should not—tackle such features. Instead, the appeal is to the technically inter-

---

13   As was already mentioned, APOLLO is a tool with which the user can create a file access structure independently of the actual file positions in the file hierarchy.

ested and experienced user. This is explicitly stated in the WINDOWS documentation, for example. Another indication of accuracy of this assessment is the fact that documentation on start-up files is frequently hived off into appendices.

Specifically in the case of WINDOWS and WINDOWS applications, however, it must be said that, as versions change, accessibility to certain parameters in the start-up files has been improved. This is true, for example, of the automatic opening of files, which is dealt with in the following paragraph. In WINDOWS 3.0, such files had to be entered manually at a certain position in WIN.INI; in the new Version 3.1, a dedicated folder exists for this purpose.

One very special, simple use of a start-up file involves merely opening it. In this case, it is often not the name that identifies a file as an "autoOpen" file, but its presence in a certain directory. This, then, holds true for all files in such a directory. The technique of opening, as a special form of evaluation, is found both in system software and in applications. Examples of the former are the Macintosh operating system Version 7, and WINDOWS Version 3.1; here, all files contained in a particular folder are automatically opened in the course of system start-up. As we know, this initiates the start-up of the application concerned, possibly along with the assigned document. So, as soon as he switches on the computer, the user is ushered into a specific application.[14]

EXCEL operates in the same way upward of Version 3 (for Macintosh and WINDOWS): With the exception of one special type, all files available in a certain folder are opened. The advantage lies not so much in the laborsaving involved, but in the fact that the execution of AUTOEXEC macros can be coupled to the opening of files. As was explained earlier, the EXCEL macro language can be used to create macros that permit far-reaching task- and user-specific adaptation of functionality and interface. If these macros are stored in *add-in macro sheets*, they, and the macro sheet itself, are not directly visible to the user and are not listed in the macro list of the macro menus[15] . All the same, they can produce their effects when the file is opened. The invisibility of add-in macro sheets prevents them, to a certain degree, from being closed and, as a result, from invalidating the adaptations performed. Similar adaptation options in combination with the start-up of an application, as described here for EXCEL, also exist in QUATTRO PRO and WINGZ. Automatic adapting of an application during start-up is important, too, because it can be combined with the possibility of automatically starting applications itself, as described in the previous paragraph. So, as an overall result, the user can be presented the

---

[14] In order to achieve this in the case of WINDOWS, the AUTOEXEC.BAT file must have an entry for automatically launching WINDOWS itself.

[15] Any shortcuts agreed to are effective, however.

adapted form of the required application right when he switches on his computer.

4. Use of the Adaptability Options

4.1. Empirical Findings So Far

Empirical studies of adaptable user interfaces usually involve prototypes that have been tested under laboratory conditions: The complexity of the application has been reduced, and the authentic work context ignored. One reason for this is certainly that adaptable systems with extensive adaptation capabilities are recent developments, so little experience is yet available on adaptability in a real work context.

Research in the laboratory gives mixed results. Studies by both Raum (1984) and Ackermann (1986) show that individual working methods can be at least as efficient as prescribed procedures and can lead to an increase in skill and in interest in the activity. Using the knowledge-based FINANZ spreadsheet system, which offers the user various design options at different levels, Rathke (1987) was able to demonstrate that an adaptable user interface can effectively support the user in performing his task. Studies of menu organization, however, have shown that menu structures individualized by the user are not significantly superior, in terms of access time and error rate, to sets of menu items grouped according to their similarity or frequency of joint occurrence (McDonald et al. 1988).

As far as real-world studies are concerned, again there is evidence of great diversity in the use of adaptation features. Koller and Ziegler (1989), studying users working with an experimental graphics editor that supported alternative input techniques for identical functionality, found considerable variety among user preferences. What is more, these preferences revealed great stability. By contrast, investigations carried out by Rosson (1984a, 1984b) during work with a word-processing editor showed that a number of users made no use of adaptation options. Even macros that were extensions to general editing functions, rather than specialized functions for experts, were not incorporated into their use of the system. Because degree of use was not necessarily associated with degree of expertise, Rosson concluded that other factors, such as the individuals' assessment of the effort needed to grasp new functions, or motivational factors, like the wish to "get something done, rather than learn more about the system", must have affected the results.

4.2. Our Own Studies of the Use of Adaptable Features

Our own examination of the use made of adaptation options did not include all the applications discussed in previous sections, because it pursued two specific goals that did not require full coverage. The first object was to examine whether users recognize existing adaptation options and incorporate them usefully into their job context, and if they do not, whether this might indicate a need to support adaptability by providing *adaptive* features. The second goal was to examine how adaptation features are presented, and whether observation of actual operations might furnish approaches for improving presentation and ways of dealing with the adaptation options. Thus, the investigation was directed functionally toward identifying adaptive and adaptability features and their design. It ran parallel to the major study discussed in this book, from 1989 to 1992.

Our aim was to find out whether, and to what extent, adaptations are used, which factors and conditions are crucial, and what benefits and drawbacks are involved in using adaptable systems. Also, we hoped to come up with possible new approaches to adaptation options on the basis of our findings.

The study of the actual use made of adaptation options was based on the assumption that user interfaces are only modified by the user where the functions for making alterations are readily apparent (discovery friendly) and easily accessible (requiring little effort) and offer a recognizable advantage (are effective) for performance of the user's task. We expected users to follow a rationalistic rather than a hedonistic approach: that it would not be the play instinct or the excitement of discovery, but cost-benefit considerations that lead to a change in the interface; and that any change would be made as an adjustment to plans of action permitting faster or easier handling of the job at hand.

4.3. Planning the Tests

In line with this approach, the point of departure for the investigation involved users' real tasks and the options available for tackling them using the given systems. The object was to observe and question users performing specific tasks using systems that could be adapted to the needs of different tasks and procedures. For each user, we wanted to identify which adaptations offered by the system would be useful in the workplace, and whether or not the user actually made use of these options. Questions regarding what benefit the users expected and actually received from the system's adaptation formed part of a questionnaire the users were administered. The study also examined whether certain changes made by the experimenters to the interface of a graphics system were noticed by users and whether they were regarded as a help or a hindrance.

We changed the names of "generic" functions in the menu to names in other systems used by the users on the same computer, and made some arbitrary changes to names of system-specific (nongeneric) menu entries.

4.3.1. Systems

In the area of office applications, some systems have user interfaces that offer a number of options for adapting both to the task and to individual user preferences. Five such systems were the subjects of this investigation:
- The word processing system Microsoft WORD 3.01 and 4.0 for Apple Macintosh (referred to as Mac-Word).
- The word processing system Microsoft WORD 4.0 (referred to as PC-Word) and WORD PERFECT for DOS.
- The spreadsheet program MULTIPLAN 3.02 for DOS.
- The database system ADIMENS for DOS.
- The graphics system MACDRAW II 1.1 for Apple Macintosh.

4.3.2. Users

Altogether, 65 users took part in the investigation.
- Mac-Word: 25
- PC-Word: 20
- MULTIPLAN: 7
- ADIMENS :   2
- MACDRAW: 11

The users were mainly scientific and administrative staff at the German National Research Center for Computer Science (GMD), about one third of whom had specific data-processing knowledge, the others having qualifications in the task domain. None of the subjects were involved with the questions underlying the project. Most of the GMD-external users worked with the software concerned on a freelance basis.

With few exceptions, none of the interviewees had ever adapted ready-made systems by directly intervening in a system or made use of any "user programming" options. They were end-users without a system development tradition .

4.3.3. Method

The experimenters had comprehensive knowledge of the adaptation capabilities of each system, as well as the standard settings of the system parameters in their unadapted form. In order to have an authentic work context, the investigations were workplace examinations in which users were questioned about the tasks they usually performed with the system and about the procedures they employed. The study involved the following steps:

(a) Identification of the adaptations made by the user.

Our first step was to examine the adaptations made by each user to the application systems concerned. First, we brought up a blank document in order to identify global adaptations, that is, those that are valid for all documents and can only be revoked by explicit reconciliation with the system's standard settings. Second, we examined the user's typical documents to find out both the adaptations that had actually been made by the user and those that could be useful for the user but had not been made.

(b) Questioning users.

All users were administered a questionnaire on the tasks at hand and on his qualifications, his familiarity with the system, and his use and assessment of adaptations, in order to try to pinpoint the criteria that determined the use or nonuse of the available options.

Additionally, a subgroup of 20 users filled out an extended questionnaire about general design options at the workplace; about individual room for maneuver, work procedures, tasks and work content, about qualifications, background experience, and information; about system use and assessment of adaptations; and about their motivation and willingness to make adaptations.

(c) Working on a task.

To underpin the data gained in the first two stages and to control the variable "task" as a relevant determinant for the use of adaptations, the test with WORD was prepared by a sample task. On the basis of a given pattern, the users had to fit three sections of unformatted text into a previously formatted target text. The task, which averaged about 15 minutes to complete, involved certain step repetitions in order to provoke the users into making adaptations.

4.4. Results[16]

4.4.1. Use of Adaptations

The vast majority of users stated that they had no trouble performing their tasks and meeting every requirement, although difficulties did arise now and then (this was also true of those few who had already adapted other systems or even done some programming).

Users made adaptations to functions in the Section, Document, and Preferences menus of word-processing systems, mainly as adjustments to task- or taste-specific features in order to control presentation factors, such as measurements: for example, converting inches to centimeters, (*customization*, as an adaptation of presentation-related parameters to customer needs). This also included styles that were, in fact, the most frequently used of all options available. These were defined individually by each user (usually 10 styles at most) and were, in general, widely used. Although users extended and modified the styles they found useful, they also defined superfluous styles or kept styles that were outdated. Nearly all users defined styles for specific documents and made them available in all of their documents.

Almost no one amended or redefined commands, for example, by defining parameters or changing key shortcuts, and only very few extended the command set (current functionality) by the addition of macros. All the same, many of the interviewees did use the available command shortcuts, especially for the generic commands like "Cut", "Copy", "Insert", "Print," and "Layout control".

Users did define macros, where that option existed (i.e., in the spreadsheet and one word-processing system; usually some two to four). They also made modifications to the menus. Although menus were extended, modified, and even abridged in one case (for the sake of a less cluttered screen, by an interviewee who described himself as a naive user), such changes rarely numbered more than one or two. Experimental adaptations of generic menu names carried out by the experimenters to reduce inconsistencies between applications were not noticed by the users and were neither a help nor a hindrance in their finding the desired function in different applications.

About one half of the users stated that they had tried out the adaptation options, if only from curiosity or interest. On occasion, adaptations were made and then forgotten, for example, when they had been defined some time previously. Also, users who had defined very many styles were unable to distin-

---

[16] Besides the authors, Cornelia Karger, Wolfgang Altenburg, Cosima Kurp, and Michael Paetau took part in the surveys. Accounts of the results have already been published by Karger and Oppermann (1991) and by Altenburg (1992).

guish between them, so they had to keep looking them up first to check which style did what.

4.4.2. Users' Assessments of the Adaptations

Utility

Most interviewees felt that the adaptations were relatively easy to make, but some did point to the importance of the time component (i.e., adaptations were difficult to make in the early stages of using an application, but became easier with increasing practice, growing familiarity, and frequency of use). Users stressed that the system itself hardly ever drew attention to the options available (at best, indirectly via the menu bar).

The interviewees were unanimous in their very positive assessment of the adaptation options. They laid emphasis again and again on easier work (greater manageability, uncluttered presentation, better work organization, and better structuring), on time saving (once the adaptations had been made), and on the lower concentration levels needed. One interviewee specifically mentioned the possibility of adaptation to his own working habits (i.e., to his personal work style). Also, some found it simply "fun" to exploit the adaptation options, and said that this added variety to their work.

So, adaptations were not necessarily made only from rational considerations of utility. Some users explored the adaptation options simply out of curiosity and interest. Of course, any changes not related to the task or the person do tend to be forgotten, and as was already mentioned, users who had defined very many styles, for example, found they could not keep them apart, so that they always had to check the style they were working with.

Adaptations that were acclaimed by most of the users and that were based on rational considerations, rather than curiosity, can be described by characteristics of the task or by the system and organizational aspects. Some tasks can benefit from adaptations both to the task procedure and to the presentation of results. The procedure might include routine, repetitive actions that can be improved by a system adaptation (for instance, by defining a macro); the task might require definition of user-specific dimensions or formats. The organizational aspect is determined by the degree of cooperation between users or groups. If users individualize their systems, cooperation can be affected by the adaptation. Another organizational aspect is the social support the user can get for adaptations from colleagues or consultants. One of the crucial aspects for using the adaptation features is defined by the system's adaptation tools. The users assess the utility of an adaptation in terms of the mental and operative effort required to discover and implement the adaptation and the self-evidence of the adaptation possibilities.

Effort

Four categories of effort can be distinguished:

- Effort needed for the task (i.e., for using the system without adaptations).
- Effort needed to learn the concept and technique of making adaptations.
- Effort needed to execute the adaptation.
- Effort needed to memorize the performed adaptations and their potential side effects.

In general, the effort needed to exploit the potential of the adaptations was said to be too high. In Multiplan, most of the work needed to produce tables is manual anyway. In word processing, too, casual users in particular, preferred manual formatting and using defaults to working with individually defined styles.

The assessment of such effort depends on the task and the benefits, and on the users' interest and knowledge. We must assume that there is an individual tradeoff between desired goals and potential drawbacks and that this determines the use made of adaptations.

Transparency

The interviewees generally found the transparency of the adaptation options and their effects unsatisfactory. In particular, the lack of transparency in functionality, when it came to the "what" and the "where" of adaptations, led to considerable difficulties.

4.4.3. Familiarization With Adaptations

The problems of transparency must be discussed in conjunction with the question of learning. The users stated that they were primarily motivated to acquire knowledge about adaptation possibilities by hints from other users and that they mostly learned about the adaptations by trial and error: Courses rarely deal with adaptation options. In a subordinate role, we find the manual as a source of help. For details about performing adaptations, other people serve as providers of assistance in the breaking-in stage or of help in the case of special difficulties. In addition to the acquisition of knowledge at the start of system use, other problems are involved in the dynamics of use, especially in modifying or extending features like macros or styles. These are either defined once and modified or regenerated during use with a specific task, or defined and continued as potentially inefficient adaptations. In extreme cases, performing adaptations may even involve making adaptations for their own sake (play-, not task-oriented adaptations).

4.5. Discussion

In spite of the use made of customization options mentioned by many users (especially in the word-processing area), it must be said that use of adaptation features to tailor a system to user needs is not, on the whole, widespread. This is also true where the task itself seems to call for the use of adaptations for the sake of efficiency.

The interdependence of features and the multidimensionality of their effects certainly allow no precise conclusions to be drawn about the factors that affect the use of adaptations. The specific difficulty is that, on the one hand, a variety of correlating variables–such as the task, the system and the user's skills or personality as independent variables–affect the use of adaptations, while, on the other hand, the use of adaptations is itself the independent variable for a number of dependent variables–such as efficiency, control of action, or user satisfaction–which also have to be considered. One possibility for dealing with these dependencies is to conduct experiments under controlled laboratory conditions on the basis of systematic variation and a corresponding reduction in variables. The limited informational value of the results–due to the displacement from an authentic work context–is only one of the problems that must then be dealt with. (Dürholt et al., 1983, referred to this methodological dilemma in analyzing complex work systems.) The advantage of workplace observation is that it enables real work contexts to be considered, such as was partly implemented in this study by an analysis of the tools. Precisely in this field of software individualization, where little recourse can be had to empirical knowledge, we must start by taking a global view of interrelated effects in order to ensure that potential additions to knowledge are not blocked by abstracting from real conditions at too early a stage. This can help us pick up initial indicators and, finally, formulate the pertinent and specific questions that are needed if an experimental analysis is to have a point at all.

This interdependency of factors made it difficult for us to determine the weight of those, influencing the use of adaptation options. For example, the use made of adaptation potential and system functions appeared to be independent of professional *qualifications* and courses attended. We found data-processing experts who made no use of adaptations, and, at the other end of the scale, users with no training in data processing who made intensive use of adaptations. Seminars and courses generally only dealt with system functions at a simple level, and there was seldom any follow-up training, so that all users more or less had to find other ways of learning how to handle the system.[17]

As far as the *task* is concerned, we observed that users performing the same task made different use of adaptations. For instance, amendments to menus or redefinitions of command shortcuts for menu options were rare, but those that were made involved considerable intervention–like extending or abridging menus (beyond simply switching from short menus to full menus). It can also be demonstrated that, in most cases, adaptations would have been conducive to work efficiency.

We also found no definite relationship between the *complexity* of the adaptations and the frequency with which they were made. Systems in which carrying out adaptations tends to be regarded as difficult were not adapted to any significantly lesser extent than those that were considered easy.

Factors, such as personality or specific problem-solving processes, were expected to contribute to the explanation of using adaptations. However, either the subjects of the study represent persons with a bias to cling to accustomed behavior rather than acquire new knowledge that goes beyond the task at hand, or the adaptation potential of the systems does not evoke a tendency to design user- and taskspecific work systems. This agrees with the findings of Rosson (1984a) and Carroll & Rosson (1987), who showed that users do not necessarily go out of their way to become experts in the use of their system, but want to get their task done. This being so, users ought to be guided and supported to adopt an explorative approach, another indispensable prerequisite for adaptable systems.

All users regarded *mental* and *operative effort* as crucial to the use made of adaptations. This suggests a situational weighing of the merits and drawbacks of using adaptations as against defaults. If, on the other hand, we also include the positive response of users who, in the study, were shown deficiencies and potentially sensible adaptations and had the execution of such adaptations demonstrated to them, we come to the conclusion that access to adaptations can be eased by careful design, so that this initial hurdle can be lowered. (Preparing for the adaptation by offering proposals, as described in the chapter by Krogsæter, Oppermann, & Thomas, this volume, is due, in part, to this assessment.)

Software that can be individualized, therefore, is by itself no guarantee of optimal system use that might meet economic or human criteria or both. A

---

[17] In an associated parallel study by Hornig (1990), again only a small proportion of seminars were found to have follow-up training schemes, although such schemes turned out to be very effective in taking into account workplace requirements and adaptation potential.

crucial part is also played by the socio-organizational environment, like the time needed to become familiar with a system, or a field of operation that offers an incentive for seeking more room to maneuver, and the set of interests of each specific user, who should be able to look on his equipment not just as an instrument, but as a part of a work environment that can be shaped (see Altenburg 1992). Another important consideration in the organizational field is the cooperative aspect of the adaptations made. Problems emerge, first, when several people have to work alternately at one PC, because they may have no record of what adaptations have been made and, second, because the more widely the various adaptation options used are, the more difficult it is to consult colleagues and ask for their assistance, because the systems then differ more and more externally (e.g., by having different interfaces). As for the system, the crucial factor is that the user be made aware of the access to adaptations, of the effort needed to execute adaptations, and of the relative relief they bring for his work, as a basis for appropriate cost-benefit calculations.

Access can be facilitated by explicitly offering "Adaptations" as a menu option, by furnishing action-relevant Help messages, or by providing access to adaptability in the form of adaptive support (see Krogsæter, Oppermann, & Thomas, this volume). In the latter case, the system might take the initiative and suggest adaptation options or demonstrate specific adaptations during suboptimal working with the standard version.

At the operational level, it appears that it might be useful to provide for the post of a software coordinator (Zink 1987, pp. 45f.), who would be in charge of all application questions (consultation, system setup, compatibility issues, etc.) or, at least, the post of a formal adviser. Such an adviser, who already exists in some firms, could be a direct contact for all users on any questions of system use, which would make it much easier for many users to approach or access a task-oriented use of the system (and its adaptation options). Problems can emerge if the various adaptations differ to such an extent that it is difficult even for formal advisers to grasp and understand the various systems employed by individual users (yielding problems of cooperation). In spite of the questions and problems described here, there is a definite trend in the technical area and at the operational level toward the forms described. It should be combined with support by adaptive features, such as those described by Krogsæter, Oppermann, & Thomas (this volume).

References

Ackermann, D. (1986):
A pilot study on the effects of individualization in man–computer interaction. *Proceedings of the 2nd IFAC/IFIP/IFORS/IEA Conference on Analysis, Design and Evaluation of Man–machine Systems,* Varese 1985. London: Pergamon Press, pp. 293–297.

Ackermann, D., & Ulich, E. (1987):
The chances of individualization in human–computer interaction and its consequences. In: M. Frese, E. Ulich, & W. Dzida (Eds.), *Psychological issues of human–computer interaction in the work place.* Amsterdam: Elsevier Science Publishers, pp. 131–145.

Altenburg, W. (1992):
Nutzung adaptierbarer Gestaltungsmöglichkeiten von Bürosystemen durch den Endbenutzer. Universität Bonn: Institut for Sociology (Diploma thesis).

Carroll, J. M., & Rosson, M. B. (1987):
Paradox of the active user. In: J. M. Carroll (Ed.), *Interfacing thought.* Cambridge, MA: MIT Press, pp. 80–111.

Dürholt, E., Facaoaru, C., Frieling, E., Kannheiser, W., & Wöcherl, H. (1983):
Qualitative Arbeitsanalyse. Frankfurt/New York: Campus.

Grob, R. (1985):
Flexibilität in der Fertigung. Berlin: Springer.

Hoffmann, S. (1988):
Adaptierbarkeit und Adaptivität von Hilfesystemen. Universität Hamburg: Institut for Informatics (Diploma thesis).

Hornig, U. (1990):
Qualifizierung zur autonomen und kompetenten Benutzung von EDV-Systemen. Universität Bonn: Institut for Psychology (Diploma thesis).

Karger, C., & Oppermann, R. (1991):
Empirische Nutzungsuntersuchung adaptierbarer Schnittstelleneigenschaften. In: D. Ackermann and E. Ulich (Eds.): *Software-Ergonomie '91. Benutzerorientierte Software-Entwicklung.* Stuttgart: Teubner, pp. 272–280.

Koller, F., & Ziegler, J. (1989):
Benutzerpräferenzen bei alternativen Eingabetechniken. In: S. Maaß & H. Oberquelle (Eds.), *Software-Ergonomie '89,* Stuttgart: Teubner, pp. 304–312.

Krause, J. (1988):
Adaptierbarkeit, Adaptivität, Intervenierbarkeit und Hilfesysteme. Universität Regensburg (unpublished report).

Kubicek, H. (1979):
Interessenberücksichtigung beim Technikeinsatz im Büro- und Verwaltungsbereich. München: Oldenbourg.

McDonald, J. E., Dayton, T., & McDonald, D.R. (1988):
Adapting menu layout to tasks. *International Journal of Man–machine Studies, 28*, 417–435.

Rathke, C. (1987):
Adaptierbare Benutzerschnittstellen. In: W. Schönpflug & M. Wittstock (Eds.), *Software-Ergonomie '87.* Stuttgart: Teubner, pp. 121–135.

Raum, H. (1984):
Aufgabenabhängige Gestaltung des Informationsangebots bei Bildschirmarbeit. *Schweizerische Zeitschrift für Psychologie, 43*, 25–33.

Rosson, M. B. (1984a):
The role of experience in editing. In: B. Shackel (Ed.), *INTERACT '84: Proceedings of the First IFIP Conference on Human–Computer Interaction.* Amsterdam: Elsevier Science Publishers, pp. 45–50.

Rosson, M. B. (1984b):
Effects of experience on learning, using, and evaluating a text-editor. *Human Factors, 26*, 463-475.

Spinas, P., Troy, N., & Ulich, E. (1983):
Leitfaden zur Einführung und Gestaltung von Arbeit mit Bildschirmsystemen. München: CW-Publikationen/Verlag Industrielle Organisation.

Triebe, J. K. (1980):
Untersuchungen zum Lernprozess während des Erwerbs der Grundqualifikation (Montage eines kompletten Motors). In: *Arbeits- und sozialpsychologische Untersuchungen von Arbeitsstrukturen im Bereich der Aggregatefertigung der Volkswagenwerk AG*, Bd 3. Bonn.

Ulich, E. (1978):
Über das Prinzip der differentiellen Arbeitsgestaltung. *Industrielle Organisation, 47*, 566–568.

Ulich, E. (1987):
Zur Frage der Individualisierung von Arbeitstätigkeiten unter besonderer Berücksichtigung der Mensch-Computer-Interaktion. Psychologie und Praxis. *Zeitschrift für Arbeits- und Organisationspsychologie, 31*, 86–93.

Zapf, D. (1990):
Möglichkeiten und Nutzung von individuellen Systemanpassungen. (Thesen zum SAGA-Workshop) Sankt Augustin: GMD (Technical Report.).

Zink, K. (Ed.), (1987):
   Arbeitswissenschaftliche Aspekte einer benutzerfreundlichen und wirt-schaftlichen Softwareproduktion. Hallbergmoos: AIT Verlag.

Zülch, G., & Starringer, M. (1984):
   Differentielle Arbeitsgestaltung in der Fertigung für elektronische Flachbaugruppen. *Zeitschrift für Arbeitswissenschaft, 38.* 211–216.

Chapter 2
Adaptivity: System-Initiated Individualization

Mette Krogsæter and Christoph G. Thomas

1 . Introduction

In recent years, the power of software products has grown immensely. The resulting complexity of systems overstresses the capacity of users to handle the available functionality. Unfortunately, in most cases the user receives no significant support in dealing with the complexity of the application.

To give a user better support, the system must be able to analyze how the user interacts with the application, and recognize when there is a problem. Unfortunately, systems seldom have built-in functions for this kind of evaluation. In order to reach the goal of more user-friendly systems, entirely new systems must be developed, or existing systems must be extended, to have the ability to analyze each user's interaction and to offer individual support.

Building systems that adapt to the user's needs is an issue both for the human factors engineer and for the system developer. They must have an understanding about adaptive system behavior in order to produce a useful and user-friendly system. The system developer is concerned with producing a successful problem-solving system, whereas the human factors engineer focuses on the development of tools that users can understand and use effectively.

In this chapter, we discuss some aspects of adaptive systems. We first define what adaptive systems are and discuss why there is a need for such systems. Then we describe what kinds of assistance an adaptive system may offer, illustrating this with references to existing prototypes. Finally, some design principles for implementing different kinds of adaptive support services are described. We discuss different kinds of modeling, along with appropriate techniques for designing adaptive system components.

2 . Flexible Systems

In the last few years, much work has been done in making systems more flexible. A flexible system increases the number of ways in which the system can be used; improves the correspondence among user, task, and system characteristics; and increases the user's efficiency. Two kinds of flexible system are of special interest: adaptable and adaptive systems.

The designer of a system with complex functionality has to make compromises in order to satisfy all possible needs of all users. In an adaptable system, the end-user may override these compromises and tailor the system as he likes. Unfortunately, adaptable systems thereby partly transfer the problems related to designing a comfortable interface from the system developer to the end user. As pointed out in an empirical study described by Jørgensen and Sauer (1990), customization features, such as tools for adapting the user interface, are hardly ever used by novice users and are used only to a limited degree by experienced users. There is a need for more intelligent system support, where the system knows more about its own tools and about the user, his work styles, and his tasks.

The idea behind any adaptive system is straightforward: The system should adapt to the user, rather than having the user adapt to the system. This basic difference between adaptive and nonadaptive systems is illustrated in Figs. 2.1 and 2.2. A consequence of making a system adaptive is that the user's frustration should disappear (a central goal of human-centered system design) or (in an ironic sense) be transferred to the computer system.

Figure 2.1:   A user faces a traditional nonadaptive system.

3. Services Offered by Adaptive Systems

An adaptive system must be able to match particular system responses with knowledge the system possesses about users and their tasks. The type of system response or adaptation can be used as one criterion for classifying adaptive systems:

• Adaptive help systems:
  These are systems that adapt the help facilities to either the task context or to specific user preferences.

- Adaptive user interfaces:
  These are systems that adapt the interaction style to the user's needs and preferences.
- Adaptive applications:
  These are systems that adapt functionality or the working of internal system modules, depending on the current state space or the user's actions.

Figure 2.2:   A user faces an adaptive system

Additionally, intelligent tutoring systems are adaptive systems that we will not single out categories on their own. These systems can illustrate aspects of help systems, adaptive user interfaces, and adaptive applications. Because intelligent tutoring systems are not our primary concern, we refer to them only when they illustrate some important aspects of adaptive systems.

Adaptive applications have not been extensively explored in the context of the SAGA project. To give you an idea of what an adaptive application can be, consider the example of an early fault-detection system for a complex process, for instance, in a power plant. When such a system detects an alarming situation, the entire functionality must change, not only the interface or help messages, in order to make it possible for the operator to rectify the situation.

Adaptive applications can be interesting from the human–computer interaction point of view, but they also raise quite a number of other questions not relevant for this book. Hence, we concentrate here on adaptive help systems and on adaptive user interfaces. Some of the systems described in the following sections could be assigned to several categories. We describe different kinds of services offered by adaptive systems, illustrating them with references to well-known prototypes.

The chapter in this book by Krogsæter, Oppermann, and Thomas describes the development of an adaptive interface, and the chapter by Fox, Grunst, and Quast describes the development of an adaptive help system.

3.1. Adaptive Help Systems

The purpose of help systems is to assist the user in problem situations. Help systems can be significantly enhanced by adapting to the individual user, because users differ in the type and amount of help that need. Each user has particular skills, experiences with the system, knowledge of the task, and so on.

The domain of help systems ranges from simple on-line manuals to systems that actively and intelligently assist the user. Table 2.1 shows a classification of help systems (Bauer & Schwab 1986).

Adaptive help systems can be passive, active, or a combination of the two. It cannot be claimed universally that either active help systems or passive help systems are of greater help to the user. Rather, both approaches have their advantages and disadvantages:

- An active system can better inform the user about suboptimal usage of the system. If the help system is passive, the user will very likely only access the help system when he knows or suspects he is working inefficiently.
- An active system may interrupt the user at inappropriate times, whereas in a passive system, the user is in complete control of when to consult the help environment.
- In an active help system, the appropriateness of the help given is much more critical than in a passive help system. If an active help message does not address the user's problem exactly and specifically, the interruption will not be considered acceptable.

We also distinguish between *context-sensitive* and *context-independent* help systems. A context-sensitive help system is a system that gives help appropriate to the user's task context.[1] In order to do this, the system must be able to infer the user's intentions. The service that can be provided by context-sensitive help systems is typically to present a selection of the most relevant help topic(s). These help topics can be inferred from the objects currently visible on the screen, from the connections to and dependencies on other objects in the background, and from the dialogue history. One of the greatest problems users

---

[1] It should be mentioned that today some help systems claim to be context sensitive by giving the user information about (user-)selected objects on the screen. However, the set of visible objects on the screen can hardly be seen as a context if the user's task is not taken into consideration.

have with static help systems is orientation (i.e., the set of available help topics may be so large that the user cannot find what is relevant). A context-sensitive system has clear advantages over a context-independent one because it narrows down this set of help topics.

passive help: invoked by an explicit user request	active help: given when the system detects that the user needs help
context-independent help: helping with the same answer to the same question, independent of the actual context	context-sensitive help: takes the special context at the time of the help request into consideration
user-independent help: the same kind of help is given to each user regardless of individual user differences	user-sensitive help: the kind of help given is adapted to the user's needs and experiences

Table 2.1: Classification of help systems.

A user-sensitive help system takes the characteristics of the user into consideration. The service given by such a system is usually related to adapting the presentation of the help to the user's expertise, preferences and so on. A problem that can occur in such systems is that the user may feel monitored or unjustly classified—after all, who wants to be treated like an idiot?

In the following sections, we describe some research prototypes illustrating different approaches to adaptive help systems, with particular emphasis on the services they offer.

3.1.1. UC, UNIX Consultant

UC is a natural-language on-line help system for UNIX (Chin 1986, 1989, 1991) developed at the University of California. UC advises users on using the UNIX operating system. Its goal is to support the learnability of UNIX by giving help appropriate to the proficiency of the user. The user modeling component of UC, KNOME, represents both knowledge about the user and knowledge about the system. The user's experience is categorized from *novice* to *expert,* and the complexity of the system concepts are categorized from *simple* to *complex.* The identification of the user's current level of experience is done with rules that make inferences based on the complexity of the concepts he uses (i.e., the UNIX commands). UC also models its own knowledge of UNIX with meta-knowledge representing facts about the limitations of the system's own knowledge-base. UC is a passive, user-sensitive help system.

3.1.2. Activist and Passivist

Activist (Schwab 1984) and Passivist (Lemke 1984) are two related help systems. Activist is an intervening (active) help system; whereas Passivist is a demand-driven (passive) help system. Both are designed to provide assistance for a text editor. Both are contextsensitive and userindependent.

Activist monitors a user's actions and intervenes when it detects the user acting at a suboptimal level. Activist's knowledge-base consists of a user model and a set of plans about how to use the text-editing commands in an efficient way. The help strategy of Activist is under user control. All parameters that determine when the system should become active can be changed.

Passivist takes questions or requests for help and interprets them while considering the user's current context: The system tries to deduce what information the user is looking for with the help request.

3.1.3. UMFE

UMFE (User Modeling Front-End subsystem; Sleeman 1985) was implemented at the University of Stanford. It is a domain-independent implementation of a system that infers a user model by querying the user. UMFE has been used in conjunction with the intelligent tutoring system NEOMYCIN (Clancey & Letsinger 1981). It can be classified as a passive, context-independent, and user-sensitive help system.

When the user makes a request to the main system, for instance, asking for an explanation of a concept, the response from the main system is handed to UMFE before it is presented to the user. UMFE tries to formulate the response by referring to concepts that the user already knows. In the user model, the concepts already known by the user are represented. If the response includes concepts that the user does not know, these concepts are left out in the explanation. If the response includes concepts that the system does not yet know whether the user knows, UMFE asks the user if he knows these concepts before the response is presented to him.

UMFE illustrates a useful user modeling technique, but has some obvious deficiencies in the user interface. It asks the user to answer several questions with "yes" or "no", and then returns a response that lists facts rather than explaining anything. Also, the system makes no initial assumption about the user; hence, it may be necessary for the user to answer many questions before the user model is of any use.

3.2. Adaptive User Interfaces

The user interface is the part of a system responsible for getting input from the user and for presenting system output to the user. A system that adapts either of these functions to the user's task or to the characteristics or preferences of the user is an *adaptive interface*.

Here are some traditional examples to adaptations of the input part:
- The set of available commands, that is, how much of a complete set of commands is available to the user, can be modified.
- Interaction objects or styles (e.g., menu, buttons, key shortcuts, direct manipulation) can be "exchanged" depending on the user's preferences.
- Additional user macros can be made available to the user.
- The system can offer to complete a task, thereby limiting the need for the user to enter data.

Here are some examples of possible adaptations to the output part:
- The content of error messages can be modified.
- For output, different presentation styles can be used (e.g., different graphical presentations).

Multimodal and multimedia systems present a new challenge of adapting both input and output parts by using different media, depending on the user's tasks, skills, and preferences. Currently, we know of no systems that exploit these possibilities to any extent worth mentioning.

Kühme, Malinowski, & Schneider-Hufschmidt (1992) presented a useful taxonomy for classifying adaptive user interfaces in a manner similar to the dimensions introduced for adaptive help systems. Different systems are characterized by describing how four different phases of the adaptation process are allocated to the system or the user. The phases that are considered are initiation of the adaptation, proposal of alternative adaptations, decision on the action to be taken, and execution of the actual adaptation (Table 2.2).

	System	User
Initiative	•	
Proposal	•	
Decision		•
Execution	•	

Table 2.2: Tasks and agents, example configuration (Kühme et al. 1992).

Our experience in evaluating adaptable and adaptive software systems showed that users prefer to make adaptations based on shared decision making. In such a configuration, the adaptive part of the system does more of an assisting job than a managing one: It prepares support and decisions for the user, while the user is free to accept or reject any suggestions from the system. Referring to the Kühme et al. (1992) taxonomy, this implies allocating the initiative, proposal, and execution phases to the system, leaving the user in control of the decision phase. We use this taxonomy to classify the adaptive interfaces described in the following sections.

3.2.1. Monitor

Monitor (Benyon, Innocent, & Murray 1987; Benyon & Murray 1988) is an adaptive interface shell that provides a framework within which different dialogues can be presented to different users. Monitor uses user, task, and interaction models. The user model consists of a set of stereotypes, each one describing a set of human characteristics. Individual users inherit characteristics from a stereotype; the system adds individual differences if it detects deviations from the stereotype.

The task model consists of a task structure model, describing the system concepts, and a function model, which is a network representation of the dialogues. For each task there are stereotypical dialogue paths designed to match the user stereotypes. This enables Monitor to adapt by providing different dialogues to different users and classes of users. In a database query application, this could, for instance, take the form of providing a form-filling dialogue for users characterized as visualizers and a command language for users characterized as verbalizers.

The interaction model collects data about the interaction during each individual session and summarizes sessions into a user-task history. This is used for making inferences about what the user understands and about the user's knowledge of different tasks. For a classification of Monitor system see Table 2.3.

Monitor	System	User
Initiative	•	
Proposal	•	
Decision		•
Execution	•	

Table 2.3: Classification of Monitor system.

3.2.2. AID

AID (Totterdell & Cooper 1986) is an adaptive front-end to the Telecom Gould electronic mail system developed at STC Technology Ltd. in Great Britain. The system adapts the level of guidance to suit the user, it recognizes that the user is familiar with another mail system, and it contains features for tracking the user's context (i.e., making it possible for the user to interrupt a current activity in order to pursue a short-term goal).

The system contains a user model responsible for inferring the user's goals and selecting an appropriate dialogue. Goal inference is done by matching the user's actions against a set of static plans, represented as Prolog facts. The system features three different heuristics that control guidance. These heuristics can also be tailored by the user; for instance, the user can suggest that the system should be more responsive or more verbose .

An application expert module contains knowledge about the mail system. It is responsible for translating the user's task into Telecom Gould commands, and for recovering from or explaining possible errors resulting from the execution.

The system was evaluated with a limited number of subjects. Unfortunately, the user interface of the system was not of acceptable quality, and this hindered the evaluation. Most important, a number of shortcomings were identified: The relationship between the interface design and adaptation had not been sufficiently considered, and there was no feedback loop enabling the system to measure the success of the adaptive changes. Furthermore, the need for presenting the user with a conceptual model of the adaptive system was identified. For a classification of AID system see Table 2.4.

AID	System	User
Initiative	•	•
Proposal	•	
Decision	•	•
Execution	•	

Table 2.4: Classification of AID system.

3.2.3. SAUCI

SAUCI was developed at Lockheed in Palo Alto (Tyler & Treu 1989). It is a knowledge-based architecture for the design of human–computer interfaces intended to support the user-oriented principles of learnability and usability. The user may select high-level tasks and is subsequently guided by the system. Menus, prompting, parameters, and help are adapted by the system and by the user. The domain of the prototypical implementation is UNIX; the system knows about 50 of the most commonly known UNIX commands.

When a user interacts with the system, a series of interaction events is generated, each event consisting of several phases. Within each phase, a rule set is activated that updates different knowledge-bases and alters the interface. The knowledge-bases of the system are a user model featuring several variables, a target system command knowledge-base, a task knowledge-base with a representation of high-level tasks, and a file or domain relations knowledge-base.

SAUCI also has help and error handling facilities that are implemented as separate dialogue sequences. To enter the help phase, the user has to ask for help explicitly. The error handling phase is entered whenever the user has made an error in filling in arguments to a command. For a classification of SAUCI system see Table 2.5.

SAUCI	System	User
Initiative	•	
Proposal	•	
Decision	•	•
Execution	•	

Table 2.5: Classification of SAUCI system

3.2.4. X-AiD

The knowledge-based system X-AiD was developed at the GMD as a proto-type for an adaptive user interface (Hein, Kellermann & Thomas 1987; Thomas, Kellermann, & Hein 1987). It is an application framework built out of a collection of asynchronous knowledge interpreters (called *specialists*) providing a set of universal or generic operations (*universals*) and some additional programs (*services*). The main declarative components of X-AiD are the static and dynamic knowledge-bases. X-AiD was used to implement some adaptive system behavior related to the user interface: active mouse positioning (if the system assumes that the user's next dialogue step is the mouse selection of a visible object on the screen, then it automatically places the mouse on that object), offering more opportunities for direct manipulation, and making dialogues shorter (hiding dialogue boxes for unused parameter settings; Horn 1991;). For a classification of X-AiD system see Table 2.6.

X-AiD	System	User
Initiative	•	
Proposal	•	
Decision	•	
Execution	•	

Table 2.6: Classification of X-AiD system

3.2.5. Eager

Eager (Cypher 1991) is a Programming by Example System for HyperCard. The user's actions are monitored and analyzed. If Eager detects an iterative pattern in the user actions, it writes a program and then suggests that the user complete the iteration by running the program. Eager uses an interface technique called *anticipation* to demonstrate how it has generalized: When it detects a repetitive activity, it highlights menus and objects on the screen to indicate what it expects the user to do next. As the user continues to perform the tasks at hand, he will notice that the objects he is about to select have already been highlighted by Eager. When Eager is incorrect (i.e., when the user's choice does not match the highlighted object), Eager will use the actual choice to revise the program. The user can tell Eager to complete the task when it becomes apparent that Eager knows how to perform it correctly. The use of anticipation allows Eager to interfere minimally with the user's normal activities. For a classification of Eager system see Table 2.7.

Eager	System	User
Initiative	•	
Proposal	•	
Decision		•
Execution	•	•

Table 2.7: Classification of Eager system

3.2.6. CHORIS

CHORIS (Tyler et al. 1991) is a generic architecture for intelligent interfaces developed at Lockheed Artificial Intelligence Center. CHORIS has been used for developing an emergency crisis management system. The set of commands available on the screen is adapted to the user, depending on the user's oganizational function (e.g., Emergency Operations Center officer or Incident officer) and the task he is working on. Input may be multimodal, especially natural language combined with pointing.

The system encompasses several models: a user model based on the stereotype approach, a domain model containing knowledge about commands and high-level tasks, and an interface model. Further, an input–output manager integrates multimodal input into a logical representation that can be used to reason about the user's actions. A plan manager currently demands that the user indicate the task he wants to execute and assists the user in the task execution. A mechanism to infer user goals from the sequence of low-level commands has not yet been developed.

An adaptor component is responsible for modifying the interface. Each significant interface event results in a transfer from one dialogue phase to another. At each transfer, the knowledge-bases are updated, some interface features are removed, and others are added. Commands that are removed are still available in a designated menu. If the user selects a command from this menu often enough, the interface will add it automatically to the main menu. For a classification of CHORIS system see Table 2.8.

CHORIS	System	User
Initiative	•	
Proposal	•	
Decision	•	•
Execution	•	

Table 2.8: Classification of CHORIS system.

4 . Design of Adaptive Systems

Any software engineer faced with the task of designing an adaptive system has to decide on an architecture for the system. These decisions include selecting a set of system modules, allocating tasks to the modules, and specifying how these system components should communicate. If we take a quick look at existing adaptive systems, it is not possible to recognize an architecture valid for all systems. Nevertheless, even from the only common requirement, that "the system should adjust to the user in some way" we can set up some guidelines (Fig. 2.3):

- The part of the application subject to changes must be a clearly separated module with well-defined interfaces to the other modules. This is necessary because this part of the system is most likely to change during development and—of course—during a session.
- User actions must be recorded and given as input to a module responsible for system adaptivity.

In order to offer some kind of adaptive assistance, the module responsible for adaptivity has to contain static and dynamic knowledge-bases or models.

The impression gained by looking into the literature on adaptive systems is that every project seems to introduce different types of models with all kinds of different names.

So, what do we really need to model? Benyon & Murray (1988) stated that "it is the automatic adaptation of systems to the changing needs of users over time and to individual users or classes of users which we need" (p. 466). This leads us to identify the two most significant models for implementing an adaptive capability: a task model responsible for noticing the different needs a user has over time, and a user model to recognize the different needs of different users.

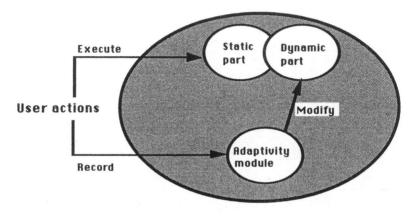

Figure 2.3:   Parts of an adaptive system.

In addition to these two models, we discuss models that can be classified as domain models and system models. A *domain model* represents knowledge from some real-world domain, and mostly serves as an information source for other system components. A *system model* represents knowledge about the system for the purpose of facilitating adaptations of functionality or the user interface. Special cases of system models are *application models* or *dialogue models*.

A system containing more than two or three models is rare. In the design of an adaptive system, it is important to identify the knowledge that needs to be modeled in order to be able to offer the kind of adaptive support in question.

When designing adaptive systems, it is useful to consider well-known techniques for the design of knowledge-based systems. Additionally, recent research on neural networks is beginning to have an impact on adaptive systems. One of the strengths of neural networks is their capacity for pattern recognition, whereas their most significant drawback is the lack of explanation capability. This suggests that neural networks may be useful for detecting unforeseen patterns in user behavior, as is also shown by two of the prototypes referred to in the section on task modeling. However, because the research within traditional knowledge-based system design seems more well-founded and general, we concentrate on design principles for knowledge-based systems. We then discuss the different kinds of models of an adaptive system and show how they can be and have been applied in order to achieve different goals.

4.1. Knowledge-Based Systems

Adaptivity represents one type of intelligent system behavior. Learning is another. Systems with intelligent behavior must be able to reason about a user's preferences and needs, his tasks, and so on. They must consider different aspects of the actual working environment and then act or react flexibly to "the user's best interest". To implement such intelligent behavior, different kinds of knowledge should be integrated into the system, especially expert knowledge and common-sense knowledge. This is best done using knowledge-based techniques, because a knowledge-based system provides a good framework for representing the required knowledge adequately and for keeping this knowledge flexible and changeable over time. Expert knowledge is knowledge about a specific domain used for a special purpose, like a problem solving process; in most cases, it is structured (for example a technical system or a description of a user interface). Common sense knowledge is general knowledge everyone may possess; it may be interpreted differently by different people (e.g., the concept "house" or notions like "good" and "bad"); it can be vague or unstructured.

Much work has been done on developing theoretical models for knowledge-based systems, and much practical experience has been gathered on how to integrate expert and commonsense knowledge into system design. Therefore, it is not surprising that the architecture of knowledge-based systems has had a great influence on the structure and development of adaptive systems. Most of the known adaptive systems use knowledge-representation techniques to build user, task, system, and domain models.

In this subsection, we give a short overview of knowledge-representation techniques insofar as they are important for the understanding of the rest of this chapter and the following chapters.

4.1.1. Knowledge Representation

In making systems more intelligent, some new issues and challenges for software and knowledge engineers arise. The engineers must be aware that knowledge must be integrated into the system, not only data; the representation, the handling, and the maintenance of intelligent system behavior must be taken into consideration. When knowledge-based techniques are used, these issues are made explicit. Some issues are summarized here to give more of a feeling for the problems (Ringland & Duce 1988). For each issue we give an example from the field of adaptive system behavior:

- Knowledge acquisition.

  The knowledge-acquisition process needs methodologies and techniques for asking experts about the domain of interest. Engineers must decide on the amount and specificity of the expert and common sense knowledge needed for this particular task. They must decide how to structure the knowledge and how deeply to model the domain to obtain the desired behavior.

  A typical example of a knowledge-acquisition problem for adaptive systems is the detection of the relevant system parameters controlling the adaptive behavior. Another example is the identification of the basic problems users have with particular applications: For example, in a help system, what is a good adaptation for the user, when does he need simple help, when complete help, and so on.

- Incompleteness.

  A knowledge-base is never complete. Therefore, engineers must identify what can be left unsaid about a domain and how incomplete knowledge can be interpreted.

  Incomplete knowledge in adaptive systems might mean, for example, that some important aspects in the user model are missing or cannot be represented. From the user's point of view, another aspect of incompleteness might be whether the knowledge-base contains all the adaptive behavior he expects.

- Reasoning efficiency.

  As in all representation problems, there is a tradeoff between expressive adequacy and reasoning efficiency. Generally, a knowledge-representation mechanism provides no means guarantee that it will be possible to perform the inference in an acceptable amount of time.

  An adaptive system, especially, lives and dies with a time-efficient interpretation of the actual knowledge. It makes no sense if a system infers an adaptation that fits to a special context, but the inference process takes so much time that the user has already left the context by the time the adaptation takes place.

The general architecture of a knowledge-based system is as follows: a (set of) static knowledge-base(s), a (set of) dynamic knowledge-base(s), and an inference engine. A static knowledge-base consists of a description of all known facts, concepts, rules, plans, and so on of the domain of interest. The dynamic knowledge-base serves as the working memory for the inference engine (or knowledge interpreter). It is a collection of the given, observed, or inferred facts about the knowledge domain (e.g., the concrete objects, the actual states, the fulfilled conditions of the rules [the facts], all recognized plans, the hypotheses, etc.).

4.1.2. Frames, Rules, Plans, and Their Interpretation

Detailed descriptions of knowledge-representation techniques can be found in Brachman & Levesque (1985) and Klahr & Waterman (1986), and in many other books and papers. In general, most knowledge-bases contain two types of knowledge: declarative and procedural knowledge. This is also true for knowledge-based systems with adaptive behavior. The declarative knowledge is highly structured. It consists of a taxonomy of objects that have attributes describing the objects and their relationships with other objects. The procedural knowledge describes the behavior of the objects in relation to each other and in concrete situations, usually in terms of rules or plans. As an example of a typical behavior that is best described declaratively, consider modeling user stereotypes or different kinds of menus in interfaces. An example of procedural knowledge might be a rule describing how a system adapts the user interface or a plan describing how a user's goal or subgoal within an application may be reached efficiently.

The concept of frames was first described by Minsky (1975). *Frames* are ways of grouping information in terms of a record of "slots" and "fillers". The essence of a frame is that it is a module of knowledge about a concept, about a situation, an event, or an object. *Slots* are places to put concrete values of knowledge about other objects, which can vary and are related to the concept. Frame systems reason about classes of objects by using stereotypical representations of knowledge, which will usually have to be modified in some way to capture the complexities of the real world (e.g., that birds can fly, but emus cannot). The properties in the higher levels of the system are fixed, but the lower levels can inherit values from higher up in the hierarchy or can be filled with specific values if the "default" fillers are known to be inappropriate.

A very natural way to represent human knowledge is to use IF–THEN rules (Cooper & Wogrin 1988). The IF part contains the conditions of the rule; it governs the premises for selecting the rule. A condition defines a pattern to be matched against the content of the dynamic knowledge-base. The THEN part contains the actions defining modifications or additions to the dynamic knowledge-base (or as it is called for rule based systems, the working memory).

If the conditions of a rule are satisfied, the THEN part of the rule is performed. This simple model describes a forward-chaining or data-driven rule interpretation. The system starts from observed data (e.g., a specific state in a user model) and proceeds to infer all possible consequences. For large rule sets, this may lead to a combinatorial explosion in the working memory. The alternative is an often-used type of rule interpretation known as *backward-chaining*, or *goal-oriented search*. Given some goal to achieve, the rule interpreter selects rules that may lead to that goal, and infers the subgoals required to satisfy those rules. The subgoals are then put into the working memory, and the cycle continues until all subgoals are satisfied.

Rule-based systems with forward-chaining rule interpretation have been successfully used to closely model human problem solving in some domains (e.g., in expert systems) or for describing adaptive system behavior. Plans or rules with backward-chaining can be used in help systems when the system tries to recognize what plan the user is trying to follow or what goal the user may be trying to fulfill.

4.2. User Models

The field of user model research is the best explored of the different kinds of models. The reason for the great interest in user modeling is obvious: If a system is to be able to adapt to the user, it must have some kind of picture of the user. The field tends to attract researchers with backgrounds ranging from psychology to computer science. This situation naturally leads to different views of what a user model really is. An *embedded user model* is a model that is incorporated into a running system and is used to increase the adaptive capability of a system. In this section , we concentrate on embedded user models.

Wahlster & Kobsa (1989) distinguished between the *user model,* as a knowledge source containing information about a specific user, and a *user modeling component* responsible for incrementally constructing the user model. In an embedded user model, both of these aspects must be considered. We see a user model as a system's model of the user's knowledge, characteristics, and preferences. We consider the user's plans and intentions to be part of the task model.

To illustrate one of the concerns of user modeling, we can take a look at different human characteristics that are believed to influence how a user interacts with a system, like intellectual ability, introversion/extroversion, field dependence/independency[2] and so on. To explore the importance of such factors, Fowler, Macaulay & Siripoksup (1987) performed experiments and concluded, among others things, that field dependence versus field independence can be usefully applied to decide which dialogue is the best one for different users. In order to classify a user on this scale, they required the users to complete a psychological test prior to the experiment, (i.e., the system did not perform any classification of the users). Benyon & Murray (1988) performed similar experiments with the Monitor system, and concluded that "at least some cognitive factors (in our case the level of working memory) can be captured unobtrusively and accurately", but at the same time, "The elicitation of people's cognitive traits is difficult. They cannot really be expected to fill in a questionnaire before using a system" (p. 471).

The examples just mentioned illustrate that human characteristics may be interesting in a user-modeling context. The problem, however, lies in deducing a user's characteristics merely from his interaction with the system. The characteristics that are modeled in current known prototypes are frequently not the most influential ones, but they are the ones that are easiest to deduce. Another problem is to map characteristics represented in a model to appropriate system responses. If such a mapping cannot be found, there is no use putting any effort into modeling the characteristic.

Modeling the user's expertise is particularly important in help systems. An example implementation is the user model in UC (Unix Consultant; Chin 1986, 1989). A user is classified as a *novice, beginner, intermediate,* or *expert,* where the measure of proficiency is knowledge of UNIX commands. This kind of modeling may prove to be sufficient when all that is needed for the adaptive component is knowledge of the user's overall expertise in using the system. However, it fails to recognize local expertise and ignores user traits other than expertise.

In order to map a user to an expertise level, UC uses a stereotype approach, a technique that can also be used for more general purposes beyond modeling expertise. A *stereotype* can be seen as a cluster of user traits that tend to appear together. The system Grundy (Rich 1979) uses the concept of stereotypes in order to present appropriate book suggestions, thereby giving an example of a typical problem area for stereotypes. Stereotyping is appropriate for matching user groups with "product" types, but fails to recognize, for instance, profi-

---

2   *Field independence* is a measure of an individual's ability to transfer knowledge from one field to another. For instance, a field-independent person is able to apply usefully his experience with text editors when learning to use a spreadsheet application.

ciency in using the system, and thus is little help for tutoring. Another problem is that the stereotype may not be fine-grained enough (i.e., the variation within the group of users mapped to the same stereotype is not taken into consideration). In the GUMS system (Kass & Finin 1988), this is taken into consideration to some extent, in that default assumptions about users within a stereotype may be modified for individual users. Chappel, Wilson, & Cahour (1992) described an extension of GUMS that allows the user model to be represented as a network rather than as a strict hierarchy. Hence, users can be classified along multiple dimensions.

A kind of user modeling frequently used for tutoring tasks involves *overlay models*, which model the user's knowledge as some subset of the domain to be learned. Hence, the model has the same structure as a the domain model, for instance, a concept tree with simple concepts being part of (or prerequisites for) more complex concepts. As a user demonstrates proficiency or a lack of proficiency with a concept, this can be marked in the user model. The course to be taken by the tutor can be found by moving upward in the tree when all underlying concepts are known or downward in the tree when the user has a problem. This kind of user modeling, also useful for modeling local expertise, is applied, for instance, in GUIDON (Clancey 1982). Another example of an overlay model is given by UMFE (Sleeman 1985). In UMFE, the relationship between concepts in the model need not be hierarchical, the implementor is free to introduce other dependencies. In the implementation of UMFE in conjunction with NEOMYCIN, concepts are arranged hierarchically, and each concept is assigned an importance and a difficulty value. This is used by UMFE for determining questions like: "Given the fact that the user knows concept c, what other concepts can the user be assumed to know?".

Another kind of modeling also known from intelligent tutoring systems is achieved by representing the user's knowledge as a set of deviations from the expert's knowledge, rather than as a subset of it. An example in which this has been applied in another context is given by the system AQUA (Quilici 1989), a UNIX help system. Here, a set of misconception classes and a set of expert beliefs are represented in different knowledge-bases. From the user's interaction with the system, the user's beliefs are determined. By referring to the knowledge-bases, the system also deduces the user's misconceptions and the beliefs that led to them. It uses this knowledge for generating natural-language responses addressing the user's incorrect beliefs.

Most approaches to user modeling found in the literature have some default model of the user to start with, or ask the user to give some information about himself in order to generate or instantiate an initial model. From that point, all changes to the model are simple modifications. There have also been attempts to design a system that can acquire a user model, so-called *implicit user modeling* (e.g., the GUMAC system; Kass & Finin 1991). The idea of implicit user modeling is to simplify the system designer's job, for instance, the huge task of analyzing groups of users and establishing a set of stereo-types. GUMAC features a set of user model aquisition rules that make infer-ences over the user's and the system's behavior and a domain knowledge-base of the actual application. The implicitly acquired model is used for advising the user.

The problem with implicit user modeling is that many sessions may be needed before the model has reached a usable state. Another problem is that the model may grow too complex and contain an information overhead (i.e., information not needed in order to produce the desired adaptive behavior).

4.3. Task Models

The term*task model* is frequently used in the literature, but is not as well de-fined as the *user model* concept. On the one hand, a task model can be seen as some kind of static representation expressing the kinds of tasks that can be performed with the system. On the other hand, it can be a dynamic representa-tion describing the task that a user is performing at a particular time. Some will view the latter dynamic task representation as an aspect of user modeling, because it expresses something about the user. We choose to view all models referring to a user's task, whether dynamic or static, as *task models*.

Tasks described in a task model will usually be user tasks; they can, how-ever, also be system tasks. In an adaptive telephone directory (Greenberg & Witten 1985), the system reacts to the frequent look-up of specific telephone numbers by reorganizing the phone number data base. This leads to faster search in future look-ups. Thus, in this example, the task model is a model of the search task, the internals of which are hidden from the user.

More frequently, a task model describes tasks performed by the user. Such a model may be a representation of different plans for the purpose of recogniz-ing user goals. An inference component tries to map user actions to actions that are parts of the represented plans. When a user goal or a set of possible and related user goals is recognized, the system can adapt in some way to make it simpler for the user to accomplish his task. If the goal can be uniquely and completely identified, the system can even offer to complete the user's task automatically. This is illustrated by the domain-independent mechanism MODIA (Schwab 1991), which provides the system designer with a language for specifying typical application system tasks. When the system is running, a task completer tries to detect partially completed tasks and, when it does, of-fers to complete the task.

Plan recognition is frequently used in help systems in order to give the user appropriate help. The problem in many nonadaptive help systems is that the user is presented with such an extensive list of help items that he does not know where to find the information he is seeking. If the space of available help items can be reduced to those that the user is most likely be looking for, the relevant information can be found more easily. If the system is able to identify the user's goal exactly (this is most likely to be the case when the system functionality is limited), the intermediate listing of possible help items can be skipped and the appropriate help given directly.

Another problem with help systems is that, in most cases, they are not invoked by users even at points where they could have proven useful. A user wants to get the job done, not to leave the context and enter a help environment. If a help system is able to recognize suboptimal or erroneous user behavior, it can give active help. For this purpose, plans expressing suboptimal task execution must be represented. Whenever the user actions can, beyond doubt, be matched with one specific suboptimal plan, the system can alert the user about a better way of reaching the goal, for instance, as it is done in Activist (Schwab 1984).

So far, we have presented some methods and applications demonstrating the power of plan recognition. Plan recognition demands that a set of plans be embedded in the system before run-time. We have not discussed the problem of plan learning (i.e., algorithms allowing a system to deduce new plans at run-time). There is one important application area for plan learning, namely, for the automatic generation of user macros.

Many systems offer macro programming languages and even macro recording facilities, making it relatively simple for the end-user to define his own macros. The definition of a macro pays if a user plans to perform a similar task repeatedly. The problem with normal macro facilities is similar to that of passive help systems:—They may be very useful, but they are seldom used. The user has to decide when to start and stop the recording of the macro—if these times are not selected properly, the macro may prove useless. The idea of automatic macro generation is that the system takes note of similar interaction sequences and offers the generation of a macro based on these "examples". The system also needs to generalize over the examples to account for slight differences among them. An example implementation is the HyperCard extension Eager (Cypher 1991), which recognizes repetitive operation sequences and offers task completion.

There are several problems with macro generation: For instance, how does the adaptive component know when it has recognized a repetitive sequence at the desired level of complexity? A user fearing to "lose control" may not want a system to take over control of very complex tasks. On the other hand, it is not helpful to offer macros for tasks that are almost as simple to perform manually.

Most task models as represented by a system designer represent either the designer's view of system tasks or are the result of extensive monitoring of users interacting with the system. Neural networks may open new possibilities for task representation, freeing the designer from some of this work. Diederich, Thümmel, & Bartels (1992) described an e-mail system containing a neural network that can be trained during usage in how to react to different e-mails depending on, for instance, the content or the originator of the message. Eberts et al. (1992) described how a neural network was trained to give relevant help to UNIX users: Different users were given the same tasks to complete; their solutions were entered as input into a neural network, along with the appropriate system response, in most cases, a help message. The advantage of neural networks for task modeling is that not all kinds of tasks must be foreseen by the system designer, because the network has a notion of "similarity". The disadvantages lie in the problem of supplying a large enough training set for the network to be stable and in the poor explanation facilities of neural networks.

4.4. Domain Models

We define a *domain model* as a representation of some domain outside the adaptive system itself for the purpose of understanding the user's reasoning, assumptions, misconceptions, and so on. The distinction between a domain model and a task model may not always be clear, especially if the user's task includes manipulating the modeled domain (for instance, process control). However, a domain model will typically concentrate on describing the domain itself, whereas a task model will concentrate on describing how to operate on the domain.

A domain model will, in most cases, be relatively static compared to a user model or a task model, in which new facts about the user or the user's tasks are continuously inferred. An exception to this is if the domain is directly connected to the changing world, as can be the case in process control systems. Another exception to this occurs if the domain knowledge of the system is not considered complete from the designer's point of view. Mechanisms must then be added for the system to be able to learn about the domain from the user.

Domain models are frequently found in intelligent tutoring systems. In this case, the domain model contains a structured description of the knowledge that is to be conveyed to the user. An example is provided by Sophie (Brown, Burton & de Kleer 1982), a tutoring system for teaching electronic circuits. Here, the domain knowledge is described by an associative net. The knowledge represented in the system is sufficient for the system to simulate circuit solutions suggested by the user and provide an evaluation based on the results.

Domain models can also be useful for providing assistance in design environments, for instance, as shown in the MMI2 system (Chappel, Wilson & Cahour 1992) for computer network design and in the JANUS-CRACK system (Fischer, Lemke & Mastaglio 1990) for kitchen design.

A simple example of a domain model in a help system is given by UC (Chin 1986), in which every Unix command is placed in one of three complexity categories. This is not in any respect a complete domain model, because no real knowledge about the commands is represented. The model is only a source of information used for determing the proficiency of a Unix user; it is not consulted for explanations. Ideally, a domain model to be used for tutoring or explanation purposes should not only be as complete as possible; the structure should also resemble structuring techniques used by humans.

4.5. System Models

A system may need to have some knowledge about itself in order to support a user. Necessary knowledge could be about its capabilities and limitations, and about its structure and functionality. This leads us to the need for a system model. A *system model* can be defined as a representation of a computer system—its architecture, interface, and so on. It is especially important if the system's functionality or user interface is the object of the adaptive behavior. If the object of adaptation is the user interface—and this is most often the case—the system model will specialize to a dialogue model, or as some prefer to call it, an *interaction model*. If the application itself is to be adapted, the system model will specialize to an application model. The model must contain knowledge about an initial system state and rules describing what changes can be made to the system.

The most widespread use of system models is within adaptive interfaces, as dialogue models describing how and when the interface should be modified. In natural-language systems, a dialogue model containing knowledge about human–computer interaction is needed in order to interpret incomplete, erroneous, or nonspecific user input dependent on the current user and the current context.

A dialogue model may also be needed to present system information to the user in a readable and understandable way. The question of how to present output to a specific user involves much more than textual/layout presentation; the user's expertise, his preferences, and which presentation medium is suitable for the message must also be considered.

For the interpretation of input and for the presentation of output alike, it is important to separate domain-independent knowledge (e.g., about natural-language parsing and general conversational principles) from domain-dependent knowledge (e.g., domain-specific terms and task knowledge).

The system Monitor (Benyon & Murray 1988) has a "task function model" that can be classified as a kind of dialogue model. Each phase of a dialogue is represented as a node with a production system responsible for how text is presented, which node is the next to be displayed, and what inferences need to be made. A further example is provided by the system SAUCI (Tyler & Treu 1988), which contains a knowledge-base that describes the functionality of the target system. Each command is represented as a separate object that contains information for constructing the interface to the target system, for instance, building menus of the system commands. Adaptation decisions, such as establishing which commands and command arguments are appropriate for a specific user, are made in conjunction with the user model.

5. Discussion

We have attempted to summarize some of the work that has been done in the area of adaptive systems. Until now, the results achieved through this work have failed to have an impact on real software systems. This can probably be attributed, in part, to the efficiency argument: An adaptive system will necessarily claim more computing resources than a nonadaptive system. However, with today's powerful computers, this should not be a serious problem.

The development of adaptive systems will necessarily take more time and, because most adaptive prototypes are either application dependent or not very powerful, it is difficult to reuse the code or concepts that have been developed for other systems. There is a need for development environments for adaptive systems. Such environments could, for instance, contain modeling shells and a tool box from which the system designer could select various adaptive services.

The developers of commercial products also consider what the market demands, and the market is not yet demanding adaptive systems. It is starting to demand adaptable systems; hence, more and more newly released products have features for user customization. The demand for adaptive systems may well spring from users being exposed to these new adaptable systems.

The need for adaptive systems will increase in the years to come (Norcio & Stanley 1989). With the increasing complexity of applications and with the introduction of new technologies, it cannot be expected that users would feel comfortable unless the system acts as a real assistant. Multimedia and multimodal systems present new challenges for adaptive systems research. Systems can be made to adapt to the user in ways still not imagined in any of the research projects we know of today. Perhaps what is most important for the success of adaptive systems in the future is to assure that HCI research does not lag behind development in other areas.

Many of the problems that remain to be solved are typical human factors issues. There must be more research on appropriate dialogues between a user and an adaptive system. Because the adaptive behavior of a system is something most users are generally unfamiliar with, care must be taken that the user does not feel dominated by some powerful adaptation component. A related problem is the fact that some users may feel monitored by the system. On the one hand, they must be given tools to inspect what the system knows about them. On the other hand, they must feel sure that this information is not being made accessible to others, something which can even have legal implications.

The kind of application involved may play a major role for the acceptance of an adaptive interface (i.e., not all kinds of adaptive behavior are suitable for all kinds of applications). The question of how to integrate an adaptive component into the overall system so that the adaptation can take place in the context of the application is also important. The adaptive component should not be isolated from the rest of the system. It must also be honest and modest, so that the user has no exaggerated expectations concerning the competence of the adaptive part: "That's one small expectation of the user, one giant challenge for adaptive systems" to paraphrase Neil Armstrong's famous description of his walk on the moon, "That's one small step for a man, one giant leap for mankind" (July 20, 1969).

References

Bauer, J., & Schwab, T. (1986, April):
Propositions on help-systems (Research Report FB-INF-86-36). WISDOM-Verbundprojekt.

Benyon, D., Innocent, P., & Murray, D. (1988):
System adaptivity and the modelling of stereotypes. In: H.J. Bullinger & B. Shackel (Eds.), *Proceedings of Human–Computer Interaction INTERACT '87.* Amsterdam: Elsevier Science Publishers, pp. 245–253.

Benyon, D., & Murray, D. (1988):
Experience with adaptive interfaces. *The computer journal, 31*, 465–473.

Brachman, R.J., & Levesque, H.J. (Eds.). (1985):
Readings in knowledge representation. Los Altos, CA: Morgan Kaufmann.

Brown, J.S., .Burton, R.R., & de Kleer, J.(1982):
Pedagogical, natural language and knowledge engineering techniques in SOPHIE I, II and III. In: D. Sleeman & J.S. Brown (Eds.): *Intelligent tutoring systems.* New York: Academic Press.

Chappel, H.R., Wilson, M.D, & Cahour, B. (1992):
Engineering user models to enhance multi-modal dialogue. In: *Proceedings IFIP WG2.7 of the Working Conference: Engineering for Human–Computer Interaction.* Amsterdam: Elsevier Science Publishers.

Chin, D.N. (1986):
User modeling in UC: the UNIX Consultant. In: M. Mantei & P. Orbeton (Eds.), *Proceedings of the CHI'86 Conference on Human Factors in Computing Systems.* Amsterdam: Elsevier Science Publishers, pp. 24–28.

Chin, D.N. (1989):
KNOME: Modeling what the user knows in UC. In: A. Kobsa & W. Wahlster (Eds.), *User models in dialog systems.* New York: Springer-Verlag.

Chin, D.N. (1991):
Intelligent interfaces as agents. In: J.W. Sullivan & S.W. Tyler (Eds.), *Intelligent user interfaces.* Reading, MA: Addison-Wesley, pp. 177–206.

Clancey, W.J. (1982):
Tutoring rules for guiding a case method dialog. In: D. Sleeman & J.S. Brown (Eds.), *Intelligent Tutoring Systems.* New York: Academic Press., pp. 201–225.

Clancey, W.J., & Letsinger R. (1981):
NEOMYCIN: Reconfiguring a rule-based expert system for application to teaching. In: *Proceedings of the 7-th International Joint Conference on Artificial Intelligence* (IJCAI '81). Los Altos, CA: Morgan Kaufmann, pp. 829–836.

Cooper, T.A., & Wogrin, N. (1988):
Rule-based programming with OPS5. San Mateo, CA: Morgan Kaufmann.

Cypher, A. (1991):
EAGER: Programming repetitive tasks by example. In: S.R. Robertson (Ed.), *Proceedings of the CHI '91 Conference on Human Factors in Computing Systems.* New York, NY: ACM, pp. 33–39.

Diederich, J., Thümmel, A., & Bartels, E. (1992):
Recurrent and feedforward networks for human–computer interaction. In: *Proceedings of ECAI '92.* pp. 206–207.

Eberts, R., Villegas, L., Phillips C., & Eberts, C. (1992):
Using neural net modelling for user assistance in HCI tasks. *International Journal of Human–Computer Interaction, 4,* 59–77.

Fischer, G., Lemke, A.C., & Mastaglio, T. (1990):
Using critics to empower users. In: J.C. Chew & J. Whiteside (Eds.), *Proceedings of CHI '90 Conference on Human Factors in Computing Systems.* New York, NY: ACM, pp. 337–347.

Fowler, C.J.H., Macaulay, L.A., & Siripoksup, S. (1987):
An evaluation of the effectiveness of the adaptive interface module (AIM) in matching dialogues to users. In: D. Diaper & R. Winder (Eds.), *People and Computers III.* Cambridge: Cambridge University Press, pp. 346–359.

Greenberg, S., &. Witten, I.H. (1985):
Adaptive personalized interfaces: A question of viability. *Behaviour & Information Technology, 4,* 31–45.

Hein, H.-W., Kellermann, G.M., & Thomas, C.G. (1987):
X-AiD: A shell for adaptive and knowledge-based human–computer interface. In: *Proceedings of the 10-th International Joint Conference on Artificial Intelligence* (IJCAI '87). Los Altos, CA: Morgan Kaufmann, pp. 97–99.

Horn, U. (1991, March):
Objektbasierte Repräsentation der Dialogschicht von Mensch-Computer-Schnittstellen als Grundlage adaptiver Systemleistungen (Tech. Rep. No. 519). Sankt Augustin: GMD.

Jørgensen, A.H., & Sauer, A. (1990):
The personal touch: A study of users' customization practice. In: *Proceedings of INTERACT '90.* Amsterdam: Elsevier Science Publishers, pp. 561–565.

Kass, R., & Finin, T. (1988):
A general user modelling facility. In: E. Soloway (Ed.), *Proceedings of CHI '88 Conference on Human Factors in Computing Systems.* New York, NY: ACM, pp. 145–150.

Kass, R., & Finin, T. (1991):
General user modeling: A facility to support intelligent interaction. In: J.W. Sullivan & S.W. Tyler (Eds.), *Intelligent User Interfaces.* Reading, MA: Addison-Wesley, pp. 111–128.

Klahr, P., &. Waterman, D.A. (Eds.). (1986):
Expert systems: Techniques, tools and applications. Reading, MA: Addison-Wesley.

Kühme, T., Dieterich, H., Malinowski U., & Schneider-Hufschmidt, M. (1992):
Approaches to adaptivity in user interface technology: Survey and taxonomy. In: *Proceedings of the IFIP WG 2.7 Working Conference: Engineering for human–computer interaction.* Amsterdam: Elsevier Science Publishers.

Lemke, A.C. (1984):
PASSIVIST: Ein passives, natürlichsprachliches Hilfesystem für den bildschirmorientierten Editor BISY. Diploma Thesis No. 293. Stuttgart: Universität Stuttgart, Institut für Informatik.

Minsky, M. (1975):
A framework for representing knowledge. In: P. Winston (Ed.), *The psychology of computer vision.* New York: McGraw-Hill.

Norcio, A.F., & Stanley, J. (1989):
Adaptive human–computer interfaces: A literature survey and perspective. *IEEE Transactions on Systems, Man, and Cybernetics, 19*, 399–408.

Quilici, A. (1989):
Detecting and responding to plan-oriented misconceptions. In: A. Kobsa & W. Wahlster (Eds.), *User Models in Dialog Systems.* New York: Springer-Verlag, pp. 108–132.

Rich, E. (1979):
User modeling via stereotypes. *Cognitive Science, 3*, 329–354.

Ringland, G.A., & Duce, D.A. (Eds.). (1988):
Approaches to knowledge representation: An introduction. Letchworth, Hertfordshire, England: Research Studies Press Ltd.

Schwab, T. (1984):
AKTIVIST: Ein aktives Hilfesystem für den bildschirmorientierten Editor BISY. Diploma Thesis No. 232. Stuttgart: Universität Stuttgart, Institut für Informatik.

Schwab, T. (1991):
A framework for modelling dialogues in interactive systems. In: H.-J. Bullinger (Ed.), *Human aspects in computing: Design and use of interactive systems and information management.* Amsterdam: Elsevier Science Publishers, pp. 940–945.

Sleeman, D. (1985):
UMFE: A user modelling front-end subsystem. *International Journal of Man–Machine Studies, 23*, 71–88.

Thomas, C.G., Kellermann, G.M., & Hein, H.-W. (1987):
X-AiD: An adaptive and knowledge-based human–computer interface. In: H.-J. Bullinger & B. Shackel (Eds.), *Proceedings of Human–computer interaction INTERACT '87.* Amsterdam: Elsevier Science Publishers, pp. 1075–1080.

Totterdell, P., &. Cooper, P. (1986):
Design and evaluation of the AID adaptive front-end to Telecom Gold. In: M.D. Harrison & A.F. Monk (Eds.), *People and computers: Designing for usability.* Cambridge: Cambridge University Press, pp. 281–295.

Tyler, S.W., & Treu, S. (1989):
An interface architecture to provide adaptive task-specific context for the user. *International Journal of Man-Machine Studies, 30*, 303–327.

Tyler, S.W., Schlossberg, J.L., Gargan Jr., R.A., Cook, L.K., & Sullivan, J.W. (1991):
An intelligent interface architecture for adaptive interaction. In: J.W. Sullivan & S.W. Tyler (Eds.), *Intelligent User Interfaces.* Reading, MA: Addison-Wesley, pp. 85–109.

Wahlster, W., & Kobsa, A. (1989):
User models in dialog systems. In: A. Kobsa & W. Wahlster (Eds.), *User models in dialog systems.* New York: Springer-Verlag, pp. 4–34.

Chapter 3
A User Interface Integrating Adaptability and Adaptivity

Mette Krogsæter, Reinhard Oppermann, and
Christoph G. Thomas

1. Introduction

To give the user the best possible support for the task and to meet specific user demands, systems should be adaptable. The adaptation may involve the positioning, presentation and naming of the system's interface elements, as well as the system's behavior, the sequence of dialogue steps, and the definition of parameter default values. The adaptation can be based on different rationales. One approach is to provide exhaustive adaptation possibilities for all interface features of the application, regardless of their potential usefulness to the user. Another possibility is to select appropriate methods and tools for adaptation and provide those adaptation possibilities that are easiest to implement. Both approaches are technology driven. They start with a product or tool kit and deliver technical features to the user, without taking the user's needs into consideration. We have used a different approach. We analyzed a given application for adaptations required to fit the application to the user's needs. The basis for the analysis was a spectrum of representative tasks selected from a commercial tutorial and from authentic work of users of the application. The application was the spreadsheet program EXCEL™,[1] which was also used for the dynamic context-sensitive help system described in the next chapter. The spreadsheet application was best suited for adaptations because it supports a wide range of tasks. The use of a spreadsheet implies a high proportion of interface interaction and routine operations of a specific user. The program indirectly provides opportunities for analyzing how the user interacts with the system and for investigating when adaptations might be appropriate by keeping an accessible record of the user's interaction. A further advantage of this existing system was that its functions could be modified. It was also possible to add functions or function accesses for specific tasks.

Adaptable user interfaces partly transfer the problem of designing the interface from the system developer to the end-user. One may argue that it is a sign of poor system design if a user feels the need to change it. The element of truth in this is that, of course, the design of the user interface is important even if the system is adaptable. However, different users have different prefer-

---

[1]   Excel from Microsoft® is a widely used spreadsheet program avaiable for a number of platforms.

ences and work styles, and one user may do different tasks at different times. The designer of a system with a complex functionality will have to make compromises in order to satisfy all possible needs. In an adaptable system, the end-user may override these compromises and tailor the system as he likes.

In systems that are solely adaptable, the user alone is responsible for when and how to adapt. On the other hand, systems that are solely adaptive may change their characteristics without any consultation with the user. Our experience in evaluating adaptable and/or adaptive software systems showed that users prefer to deal with systems based on shared decision making: The user can act on his own and gets support from the system whenever he likes. The adaptive part of the system does more of an assisting job than an executing one: It prepares support and decisions for the user, but the user is free to accept or reject the suggestions from the system.

2. Constraints on the Development and Evaluation of Adaptation Concepts

A critical point in the development of adaptation features is that the basic application system should be complex enough to justify advanced adaptation features. Until now, the development of adaptive systems has mainly been carried out in research institutions with limited resources. As a result, mainly small prototypes have been built with a complexity far from that of real applications. The problem with this kind of development is that if an application is that simple, adaptations can only cover specific aspects of the system. The evaluation of the resulting system can only take place in an arbitrary test setting. Furthermore, the criteria for the evaluation will be of limited significance (mostly errors and the time spent on different functions). We did not try to overcome these limitations by the development of a more complex application, but looked instead for a system already in use in the real world: a well-known and well-designed complex application developed by a commercial software house that we could *extend* with a flexible environment of our own. Because software houses are generally unwilling to give away source code, it seemed at first glance to be impossible to find an existing application to modify. There is, however, another way to extend closed applications, namely, through programming in the macro language that comes with the system (according to the same principle that was used for implementing context-sensitive help, as described in the chapter by Grunst, Fox, & Quast, this volume).

The spreadsheet program Excel for the Macintosh™ was found to have several features making it possible and reasonable to extend with an adaptation environment:

- The system and the user interface are complex enough for adaptivity to make sense.
- The system has an up-to-date graphical user interface into which the extensions can be naturally integrated.
- Users can be found who are familiar with the application, thus making a realistic evaluation possible.
- User actions can be recorded. It is, therefore, possible to make an on-line evaluation of what the user is doing.
- The user interface is modifiable (new dialogues, menus, etc.), facilitating the introduction of extensions.

Because of these advantages, we decided on Excel as a basis for developing the adaptive system, Flexcel (flexible Excel).

3. Design Cycle of Flexcel

We used two versions of Excel for the design of an extended application with new adaptation features. In the first experimental system of Flexcel (Flexcel I), we used Excel 2.2. In this version, Excel showed some limitations for the design of adaptations (see Section 3.2). After the evaluation of Flexcel I, some system deficiencies could be identified. We used this experience to formulate the requirements for the second experimental system, Flexcel II. At that time, the new version of Excel, Version 3.0, was released with more facilities for the development of customized applications.

3.1. Technical Considerations

Both Excel 2.2 and Excel 3.0 offer several different kinds of interaction:
- Command selection from menus in the menu bar.
- Key shortcuts for some of the commands listed in the menu bar and for several actions not listed in the menu.
- Parameter entry in dialogue boxes.
- Selection of icons from an icon bar for some frequently needed operations (in Version 3.0).
- Direct manipulation (with a mouse) of windows and spreadsheet cells.
- Entry of data and formulas using the keyboard.

Excel itself offers few adaptation possibilities to the end-user. The user may choose between long and short menus; commands that are considered complex or rarely needed are not included in the short menus. The user may also enable or disable a status bar and the icon bar (only in Version 3.0). These adaptation possibilities of Excel limit themselves to the *reduction* of available functionality or information. Additionally, the user can define his own macros by letting Excel record a series of actions or by means of a macro programming language. We wanted to make it possible for users to extend the system without having to enter a programming environment. This meant that we would have to extend Excel with new functionality. Using the Excel dialogue editor and the macro programming language, it is possible to design a completely new system:[2]

- The Excel menu bars may be modified and new menus created. Commands can be deleted, and new commands can be added. The effects of these new commands are defined by macro programming.
- For new commands, dialogue boxes, and corresponding dynamics can be defined.
- New key shortcuts can be defined.

This enabled us to design the Flexcel system to look like the normal Excel, with the adaptation capability as a natural extension. Excel has some limitations for the programmer, however: The icon bar in Version 3.0 cannot be changed, the Excel dialogue editor sets limits on the complexity of dialogue boxes, and the method for direct manipulation of spreadsheets cannot be changed or extended. One limitation was concerned with the recording of user actions. This had to be done via file transfer, which was rather awkward and slow. Also, some user actions are difficult to record, especially direct manipulation. In spite of these limitations, Excel met most of our requirements.

3.2. Flexcel Adaptation Guidelines

Before and during the project, we set up several guidelines for adaptive and adaptable systems, which were taken into consideration during the specification and implementation of Flexcel. It was necessary to formulate the guidelines explicitly, because, as we found during the initial phase of the development, it is easy to "forget" general principles. The same poor solutions tended to be suggested over and over at regular intervals. The most important guidelines were the following:

- For every adaptive capability of the system, there must be an equivalent adaptation capability available to the user.
  Justification: Users must know that they are allowed to do everything that the system can do.

---

2   These programming features are very useful for writing custom applications, especially spreadsheet applications for special purposes.

- There should be easy and flexible access to the adaptation environment. Justification: Customization features are of little use if they are difficult to access.
- At any time, the user should be in complete control of the system; the system may only act as an assistant.
  Justification: The user is expected to be creative. Therefore, the user should choose the most suitable work style.
- Suggestions from the system should not disturb the user unnecessarily in his work.
  Justification: System adaptation features are only aids to the user, and suggestions should not take the user's attention away from the real task.
- Whenever possible, more than one adaptation should be offered.
  Justification: A system is usually not able to identify the user's needs precisely. The user should have the freedom to select among a number of different adaptations suggested by the system.

3.3. Scope of the Adaptation Environment

The first thing we had to do was to define the scope of the adaptation environment. The overall goal was to develop an adaptation environment making it possible to perform realistic experiments. From these experiments, we were hoping to gain new insights into the usefulness of our approach, in particular, and about adaptivity and adaptability, in general.

Various constraints also had an impact on defining the scope, especially limitations of Excel and response time considerations. We decided to develop a modest but realistic adaptation mechanism that could be used for authentic tasks. Most of the work would be put into deciding how to present information to the user, rather than into developing elegant internal algorithms.

The initial idea for the adaptations offered by Flexcel grew out of our own experience with Excel as users. A simple scenario may illustrate the idea: When an Excel user needs to add the value of a cell to a set of entries of other cells, he must select the function Paste Special from the menu, then specify the parameters Values and Add and finally click the OK button in the dialogue box. If the user needs to do this repeatedly during a session, he will start wishing for a simpler and faster procedure. This was the origin for the following ideas for simplifications:

- A new menu entry or a key shortcut for the procedure could eliminate the dialogue step entirely.
- The selections in the dialogue box could be defined as new parameter defaults. For our example, this would mean that the dialogue box would already have the parameters Values and Add when the dialogue box for Paste Special appears on the screen.

These options were to be available to the user through an easy-to-use interface, thereby making the system adaptable. Further, an adaptive system component should present proposals for such system adaptations to the user. This adaptive component was to detect repeated function executions with identical parameters, inferring the need for a system adaptation whenever the number of identical calls exceeded a threshold value. To do this, we would have to record user actions, let a knowledge-based component analyze this log, and present the resulting suggestions to the user.

4. Flexcel I

The design of Flexcel I was not entirely specified in advance. Rather, it reflected the results of design–evaluation–redesign cycles. The process started with the adaptation ideas described in the previous section. In addition, we analyzed adaptive and adaptable concepts in existing prototypes or systems and considered opinions expressed by human factors experts (see Reiterer & Oppermann 1993). A complete report on the principles of timing and displaying adaptive suggestions and of the evaluation of the usability of Flexcel I is found in Oppermann (1991). A technically oriented system description is given in Brüning, Krogsæter, & Thomas (1991).

Flexcel I was a customized spreadsheet application that provided most of the functionality of the original Excel (slightly more than the basic version with "Short Menus"). Some of the functions of Excel were modified to allow different types of adaptations. From the analysis of tutorials and authentic tasks, we identified two groups of functions somewhat different in nature. For instance, the dialogue boxes of some functions (especially format functions) do not display predefined parameter defaults for these functions; rather, the parameters set in the dialogue box reflect the status of selected cells in the spreadsheet. This suggested that it would be unnatural to offer the opportunity to set defaults for these functions. For each function group, we decided on the adaptation types to offer:

Function Group 1:

• Given parameter defaults could be changed, or the dialogue step could be eliminated entirely by fixing defaults, making "normal" function execution (with the dialogue box) possible only by means of a key shortcut.

Function Group 2:

• New key shortcuts for different function parameterizations could be defined.

The following Excel functions were made adaptable and adaptive: Clear, Paste Special, Insert Cells, Delete Cells, Number Format, Alignment, Font, Border, Extract Data.

The adaptive feature of the system consisted of a mechanism that suggested new key shortcuts, changing parameter defaults, or eliminating the dialogue step of functions. A suggestion appeared in a dialogue box on the screen whenever the number of identical function calls had exceeded a threshold value. The analysis of user actions was done by rules in a simple *usage profile*[3] with statistics about how frequently different function parameterizations were selected by the user.

Flexcel I was evaluated in an HCI Laboratory. The setup of the lab allowed for empirical interaction analyses with users performing test tasks. The evaluation did not conform to a classical psychological experimental design. The evaluation method was selected in accordance with the question to be answered. It was a design-cycle–oriented question of "What is bad with the system and why?", rather than a question of "How good is the system?"[4]. The aim of the evaluation was to identify the cognitive model users develop in interacting with the adaptation component of the system and what problems they have in using and understanding its mechanics and behavior. The evaluation used an exploratory approach using "weak methods", such as video observations, video confrontation, and interviews (see Carroll & Campbell 1986)[5].

Users were given the task of completing a spreadsheet, during test sessions that were videotaped and analyzed. We performed the evaluation with users who already had some experience with Excel. We prepared for the actual tests with some reconnoitering runs to detect and fix obvious shortcomings and technical errors of the system. In the experiment, the users had to execute written test tasks. The experimenter acted as a communication partner for the test person. The resulting verbal interaction was useful for interpreting the problems and assumptions of the subject. The sessions were recorded on videotape and on an on-line log file. After the tests, we conducted interviews and, in some cases, discussions with video confrontations.

---

[3] We use the term *usage profile* instead of *user model* or *user profile*. This should emphasize that our main focus is to model and analyze the use of a computer system by a user and not the user himself.

[4] Answering the question, "How good is the system?" implies comparing different alternatives under certain aspects or testing a given system against fixed criteria (cf. Fähnrich & Ziegler, 1987).

[5] Experiments are characterized as "hard" methods when statistically analyzable data like time and errors, are collected (quantitative, theory driven—cf. Newell & Card, 1985; see also Oppermann, 1991).

In the case of Flexcel I, five users[6] participated in the tests. Two of the subjects were informed about the adaptation opportunities, but not about the adaptive capabilities of the system. The others approached the adaptability and the adaptivity without any prior preparation. The users were instructed to complete and format a spreadsheet and to extract classes of records from a database.

The evaluation showed a serious problem with the way adaptation suggestions were presented to the users. Suggestions appeared on the screen without warning, and the users had to respond to the suggestion by acceptance or rejection. Several users did not know how to respond in such a situation. Also, some adaptations would change the behavior of a function without this being clearly visible in the user interface: For example, a function could be executed without the dialogue step, but the parameters applied in the execution would not be shown.

The tests showed that users easily became confused by such changes when they wanted to use a changed function in a later phase of the session. The fact that users had to explicitly accept the changes was not a sufficient solution to this problem. To explain the problem, we must consider that adaptations are performed after one has learned the basic concepts and the basic behavior of the system while proceeding with routine tasks. They confront the user with problems similar to those of different system modes. The user has to remember initial and adapted versions of the system. The user must be supported by the system in handling this difference; for instance, there must be an easy way to reset system adaptations. Flexcel I was shown to have serious deficiencies in this respect.

It could also be seen that, in connection with an adaptation suggestion (especially one proposing to eliminate the dialogue step of a function), users would experience a cognitive overload. With an adaptation proposal, the user was presented with a suggestion, asked for confirmation, and informed about the new behavior of the system. Hence, the user was confronted with three kinds of cognitive demands:

- To turn from task execution to an adaptation suggestion and to devote attention to a metatask.
- To evaluate the suggested configuration of the system in light of current and future tasks and personal preferences, and to decide whether to accept the suggestion.
- To understand the different adaptations and memorize their effects.

---

6   The tests were designed to show shortcomings in the presentation of the adaptivity features. Only a few subjects were necessary to identify cognitive and operative problems and new design requirements.

This demands concentration on too many mental perspectives in a single situation. From each cognitive problem domain, we derived specific design requirements for the second version of Flexcel, Flexcel II.

5. Flexcel II

5.1. Changes From Flexcel I

Based on our experiences with Flexcel I and new extension possibilities provided by the macro language in Excel 3.0, we identified a number of new solutions for Flexcel II (see Thomas & Krogsæter 1993). The major changes from the first version can be summarized as follows:

- Access to the adaptation tools
  In Flexcel I, access to the tools was over a menu entry "Adaptations". Users had difficulties finding this entry; hence, in Flexcel II we introduced a separate adaptation *tool bar* to make the tools more visible.
- Presentation of suggestions:
  In Flexcel I, adaptation suggestions appeared in dialogue boxes without warning. In Flexcel II, we wanted to let the *user* decide when to read the suggestions. Therefore, the presence of a suggestion was indicated by an acoustic signal and a blinking button in the tool bar. Unread "tips" were maintained in a "Tip List".
- Extended rule base:
  Flexcel I had a very primitive usage profile, in which only the *execution* of the different functions was modeled. Flexcel II also modeled the use of the adaptation tools, the handling of suggestions from the system, and the user's interaction style.
- Other extensions:
  Flexcel II contains a number of other features that were not part of Flexcel I. These include the possibility of defining new menu entries, the appearance of usage suggestions reminding the user of adaptations he seems to have forgotten, and a critique module (see Fischer 1990) telling the user how the adaptation tools might be used more efficiently. The critique focuses its attention on the user's special needs and behavior concerning the adaptability. It consists of a domain knowledge-base with a set of rules and a usage profile. In our approach, the domain is "adaptability", and the usage profile describes the individual use of the adaptation possibilities. The main task of this extension is to recognize suboptimal usage of the adaptation tools.

Similar adaptation options were specified and implemented in Flexcel II as in Flexcel I, but they were designed and presented in a different way. For instance, we abandoned the strict distinction between the two groups of functions in order to reduce users' confusion that resulted from their having to handle functions differently. This meant that in Flexcel II users could define menu entries and key shortcuts for all adaptable functions. The ability to set defaults was retained for some functions, but this was done by simply adding a button to the dialogue box, which was otherwise identical for all functions. Also, the possibility of eliminating the dialogue step of a function was removed. It had proven confusing in Flexcel I, and the introduction of user-definable menu entries made it redundant.

5.2. The Flexcel Tool Bar

The Flexcel tool bar shown in Fig. 3.1 serves as the main control panel for adapting Flexcel.

Adaptive suggestions are presented as "tips" and are indicated by a tone, three blinks of the button "The Tip" in the adaptation tool bar, and a subsequent corona around the icon. In order to read the tip, the user can click the button.

Unread tips are collected in a "Tip List" so that they can be accessed by the user at any appropriate time.

The button "Adapt" presents to the user a list of all adaptable Excel functions. The functions can be selected, new adaptations can be defined, or existing adaptations can be changed by the user.

The button "Overview" displays a summary of all user-defined keystrokes and menu entries.

The "Critique" button may be clicked whenever the user wants his interaction with the adaptation environment to be critically analyzed. The user is told in what ways the adaptation tools can be used more efficiently.

Clicking on the "Tutorial" button starts a tutorial on the adaptation facilities. The tutorial is an interactive animation explaining the basic ideas of how to adapt the interface.

Figure 3.1:   Flexcel adaptation tool bar.

Flexcel provides three kinds of suggestions: adaptation tips, usage tips, and tutorial tips. *Adaptation tips* recommend that the user define a menu entry and/or a key shortcut for some function parameterization. Such a suggestion appears after the user has repeatedly executed a function with identical parameters. Figure 3.2 shows the dialogue box for an adaptation that might appear after clicking the "Tip" button in the tool bar. As can be seen, even if the system recommends defining a new menu entry, the user can instead define a new keystroke for the suggested parameterization.

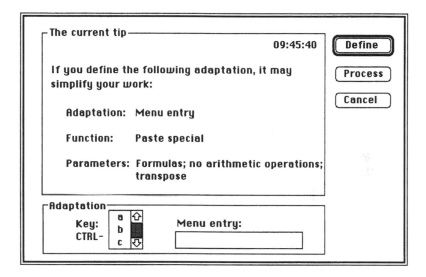

Figure 3.2: A Flexcel adaptation tip.

A *usage tip* is a reminder: It informs the user about an adaptation he has defined but seems to have forgotten. A usage tip may appear after a function has been repeatedly executed with a parameterization for which there already exists a menu entry and/or a key shortcut. An example of a usage tip is shown in Figure 3.3.

A *tutorial tip* recommends a tutorial that explains how to adapt the system. A tutorial tip appears if the user regularly accepts adaptation suggestions and uses the adaptations, but does not adapt independently.

Besides the system support for adapting the interface, the user can, at any time, adapt on his own initiative. Figure 3.4 shows the dialogue box for "Font" as it appears after the user selects this function from the menu. The dialogue box looks like the normal Excel dialogue box for this function, with the addition of a button "with adaptations". When this button is selected, the dialogue box expands to a combined execution and adaptation box, as shown in Fig. 3.5. The user can define key shortcuts and/or new menu entries for the selected parameters in the upper part of the dialogue box.

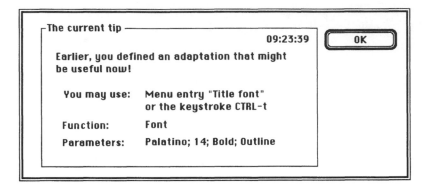

Figure 3.3:   A Flexcel usage tip.

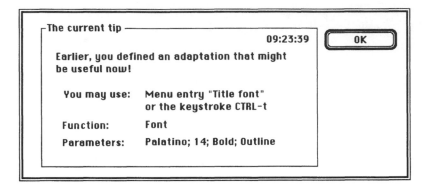

Figure 3.4:   Normal execution dialogue box for function "Font".

The adaptation part of the dialogue box in Fig. 3.5 is basically the same for all adaptable functions. The user may enter parameters in the upper part of the box, and then he may define a key shortcut and/or a menu entry for the specified parameterization. He can create definitions for several different parameter combinations without leaving the dialogue box (symbolized by the

>> on the "Define" button). The user can select an available key from a key list or directly type in a key. He can inspect a list with the actually defined entries for parameterizations within the function at hand and call up an overview of shortcut definitions of the entire system. The function "Font" does not have a set of default values; therefore, a button "Set default" is missing in this dialogue box.

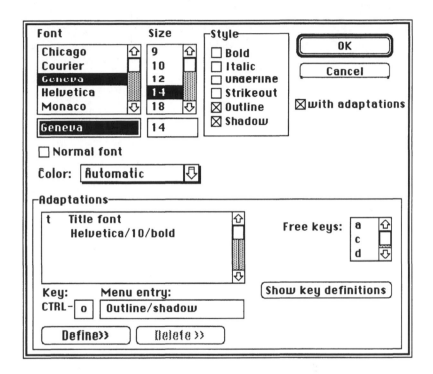

Figure 3.5: Combined adaptation/execution dialogue box for function "Font".

Thus, the adaptation environment is accessible in two different ways. First, with the button "Adaptations" in the described tool bar, and second, with a button "with adaptations" in the dialogue box of each adaptable Flexcel function.

5.3. Flexcel System Architecture

The adaptive extension Flexcel was developed on a MacIvory™: a Macintosh with a Lisp co-processor board. It is implemented using the Excel macro language and dialogue editor, as well as CLOS.[7] Excel and the user interface run on the Macintosh; the knowledge-base responsible for generating adaptation suggestions and critique runs on the Lisp Machine.

Figure 3.6 illustrates how data flows in Flexcel. Communication processes on the Excel and on the Lisp sides, shown as grey circles in the figure, control the exchange of information. All user actions are recorded in Excel and then sent to the knowledge-base of the Lisp machine. The incoming entries are LISP expressions, which need only be evaluated. Thus, the entries can be seen as events that trigger different rules of the knowledge-base. The activation of rules may lead to an adaptation tip or a usage tip, which is then transferred back to Excel, thereby activating an Excel macro which causes the "Tip" button in the tool bar to blink.

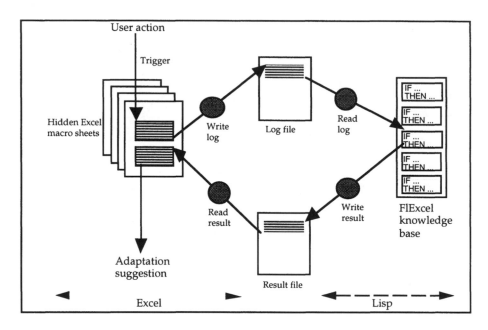

Figure 3.6:   Data flow in Flexcel.

---

[7]   Common Lisp Object System.

The Flexcel knowledge-base consists of a set of rules and a dynamic usage profile. The usage profile is a CLOS object with a set of attributes, several of which are also CLOS objects. Protocol entries are interpreted as *events* leading to changes in the usage profile. Changes in the usage profile may trigger rules, implemented as *methods* connected to the corresponding objects. In the following sections, some aspects of the Flexcel knowledge-base are described. For clarity, rules are described using a simple IF–THEN syntax.

5.3.1. The Flexcel Usage Profile

The usage profile contains all data appropriate for generating adaptation suggestions. This includes a description of the current state of the Flexcel system (e.g., all user-defined menu entries), data about the usage of different functions, and data about the user's interaction with the adaptation environment. The main attributes of the usage profile and some related rules are described in the following:

SelfAdaptPreference:
> This attribute indicates to what degree the user prefers to adapt his system environment independently (i.e., without waiting for a system suggestion).
> Rule example:

IF	the user makes an adaptation
AND	this adaptation has not been suggested by the system
THEN	increase the SelfAdaptPreference

SystemAdaptPreference:
> This attribute indicates the user's attitude toward the adaptation suggestions from the system.
> Rule example:

IF	an adaptation suggestion is accepted
THEN	increase the SystemAdaptPreference

AdaptationTipThreshold:
This attribute describes the number of times the user may execute a function with a given parameterization before an adaptation suggestion is generated. The threshold is low if the user seems to appreciate the adaptation tips; it is high if the user prefers to adapt without any system interference. The value is calculated from the attributes SelfAdaptPreference and SystemAdaptPreference whenever one of them changes:

Rule example:

IF	SelfAdaptPreference changes
OR	SystemAdaptPreference changes
THEN	update AdaptationTipThreshold

UsageTipThreshold:
> The use of this attribute is similar to that of the AdaptationTipThreshold attribute, but controls the occurrence of usage tips. However, the value of UsageTipThreshold is fixed.

UsageStatistics:
> This attribute is an object instance that contains a number of data describing the user's work style, such as the number of keystroke errors and the ratio of menu calls to key calls. The data influence the *kinds* of adaptation suggestions that are given: To a typical menu user, suggest the definition of new menu entries; to a typical keystroke specialist, suggest the definition of new key shortcuts.

> Rule example (updating of UsageStatistics):

IF	an Excel function is activated from the menu
AND	this function has call type MenuAndKey
THEN	increment NumberOfMenuCalls in UsageStatistics

Functions:
> This is a list of all functions in the Flexcel menus, each function being an instance of one of the object classes ExcelFunction, FlexcelFunction, or FlexcelFuncWithDefault, the latter two classes representing the adaptable functions. A simplified description of these object classes with attributes and inheritance is shown in Fig. 3.7. The utilization for some of the internal attributes of the functions, among others, are illustrated in the following section about rules for tip generation.

5.3.2. Rules for Tip Generation

After each execution of an adaptable Excel function, the corresponding log entry triggers rules that may result in either an adaptation tip or a usage tip. For each function, a set of tip candidates is maintained. Each candidate is an object instance consisting of a particular *parameterization* (ref. also to Fig. 3.7) and a *frequency* indicating how many times the function was executed with this parameterization.

In order to prevent the list of tip candidates from becoming unwieldly—this would make the system gradually slower during a session—the number of tip candidates is limited. Each tip candidate has a *recency* attribute, measuring how current it is. When the set of tip candidates is full and a new candidate arrives (through a function execution), the least recent candidate is discarded, and the new candidate gets the free slot.

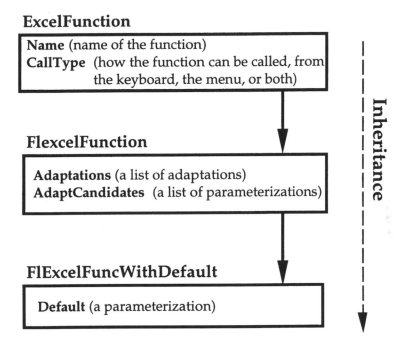

Figure 3.7: Object classes representing Flexcel functions.

Rules used for the generation of adaptation and usage tips (p = Parameterization of the execution to be evaluated) are:

> IF   there exists a tip candidate with p
>   AND   there is no adaptation with p
>   AND   the frequency of this candidate is equal to
> AdaptationTipThreshold
> THEN   Generate tip:
>   Recommended adaptation: Define a menu entry and/or a key shortcut (depending on the UsageStatistics) for p.

> IF   there exists a tip candidate with p
>   AND   there exists an adaptation with p

AND   the frequency of the tip candidate is equal to
UsageTipThreshold
THEN   Generate tip:
   There exists an adaptation (keystroke and/or menu entry) that can
be used for p.

5.3.3. Critique Rules

The critique rules of Flexcel are triggered only when the user asks for them by
means of the "Critique" button in the tool bar. The given critique is restricted
to the user's interaction with the adaptation tools. The aim of the critique
module is to give the user hints as to how the tools might be used more efficiently. If a user seems to have no problems with the tools, no criticism will
be presented. The following describes three of the possible critique types and
the corresponding rules:

A user adapts, but does not use the adaptations:

IF   the number of usage tips is high
  AND   most of the usage tips were read
  AND   one-of(SystemAdaptPreference,
SelfAdaptPreference) is high
THEN   Explain the adaptation concept, and offer the dele-
tion of unused definitions.

A user tends to accept adaptation suggestions, but does not adapt
independently:

IF   the SelfAdaptPreference is low
  AND   the SystemAdaptPreference is high
THEN   Point to the "Adaptations" button in the tool bar
and explain the normal adaptation dialogue.

A user has defined several key shortcuts, but makes many mistakes
when using them:

IF   the KeystrokeQuotient is high
  AND   the KeyErrorQuotient is high
THEN   Point to the "Overview" button, and explain how
menu entries may be defined for the already defined keystrokes.

6. Evaluation of Flexcel II[8]

Eight users took part in the actual experiments with Flexcel II. One subject had prior information about the adaptation features. The task was to edit the content and layout of a spreadsheet. Unlike the test tasks for Flexcel I, the description of the test tasks for Flexcel II was not given in procedural form. Instead, a final form of the spreadsheet to be produced was given to the users. The users could decide how to proceed in order to complete the spreadsheet. This form of description gives the user greater freedom to make use of the functionality of the system and of the facilities of the adaptation component.

The users seemed to like the unobtrusive way in which adaptation tips were presented. Also, without prior information, they would focus their attention on the blinking "Tip" button and click it to see what was behind it. As they gained experience, users sometimes noticed the presence of a suggestion, but waited until a more appropriate time to process it.

Users of Flexcel II did somewhat better than users of Flexcel I in finding out how to adapt without waiting for adaptation suggestions. The presence of an "Adapt" button in the tool bar seemed to help, but for users with no prior tutorial or explanation, even this was not self-explanatory enough.

The evaluation of Flexcel II not only showed positive aspects of acceptance of the implemented adaptation concept, but also some deficiencies, which are discussed further on.

6.1. Task Performance and Adaptation

The switch from the task to the adaptation should not be the exclusive business of the system. First, this means that the system should never force the user to interrupt his task and make an adaptation or even consider an adaptation suggestion presented by the system. Flexcel I confronted the user with a dialogue box that interrupted the current work flow. This was not well accepted. Flexcel II presented the adaptation in a separate tool bar. The presence of an adaptation suggestion was indicated by cautious hints (a tone, the "Tip" button blinking three times, and the subsequent corona around the button) that did not interfere with the user's work. The tip could be ignored or taken up at any appropriate time. Tests of this presentation showed increased user acceptance, because the users could wait until their current work allowed an interruption to switch over to the adaptation task.

---

8   Results of the evaluation were reported with examples of user comments in Oppermann (in press).

6.2. System- or User-Initiated Adaptation

If adaptation is to be not the business of only the system, then users must have their own facilities for making adaptations—not only when the system suggests them. In most cases, Flexcel's adaptation suggestions inspired users to develop their own adaptations (i.e., to transfer and generalize the adaptation suggestions). The users wanted to be able to use their own adaptation facility to overcome the dependency on the adaptive component. This tendency needs to be system supported, because users have difficulties in detecting the access to and the learning of the operation of the adaptation tools. For instance, in Flexcel I we included an adaptation check box in each dialogue box of adaptable functions, but no test user identified this button as an adaptation access. They failed to understand the adaptation idea even after they had received the adaptation suggestions from both the system and the instructor or co-user. More prominent presentations of adaptation features or explicit instructions were necessary to introduce the adaptation concept to the user. Increasingly spectacular presentations (using color, size, placement, movement) compete with the presentation of other features , however, and tend to yield an inflationary cycle. In Flexcel II, we presented several different access methods to the adaptation: the adaptation check box in the dialogue boxes, an additional adaptation button in the generic tool bar, and an auxiliary adaptation tip when the user showed acceptance and use of suggested adaptations, but did not initiate any himself. The tip proposes to the user that he adapt on his own or consult a tutorial to become familiar with the adaptation tools. The evaluation showed that the transition from adaptation suggestions from the system to adaptations by the user is better with Flexcel II than with Flexcel I, but not yet satisfying. The problem is less severe after the user has been instructed via a tutorial tip or personal communication, although even after the first self-initiated adaptation, the user has an incomplete understanding of adaptation concept.

We think that we are on the right track, but have to improve the transition from adaptation by the system to adaptation by the user, in order to locate the "locus of control" for the adaptation with the user. In Flexcel II, the user can decide if and when he will respond to adaptation suggestions, he can select an adaptation from a set of options, and he has access to his own adaptation facilities, as described.

6.3. Accuracy of Adaptation: The User's Freedom

When presenting adaptation tips, the timing and content of the proposed adaptation must be appropriate. The user has particular tasks, personal needs, and preferences that must be reflected by the proposed adaptation. This suggests that the system needs to make an in-depth analysis of its interaction with the user. In order to develop suggestions that fit well with user needs, the analysis not only has to take the current dialogue situation into account, but must also consider the usage history of the specific user. In Flexcel, we tried to consider both the user's interaction preferences (preferred use of menus or keys) and the user's adaptation preferences (acceptance/rejection of tips, how often previously made adaptations are used, ratio of system-initiated to user-initiated adaptations) according to the rule base described previously.

Despite these efforts, evaluation of the use of the different system versions seems to indicate that it is impossible to meet the exact needs of the users. Not only machines, but even human advisers have problems meeting these criteria, as observed in the Wizard of Oz settings (see also the chapter by Fox, Grunst , & Quast) in our laboratory. In these situations, users had to perform a test task, and a system expert observed the session via a system–system connection. The expert was restricted to the information received in this way: He had no oral or visual communication with the user. In problem situations, the human adviser could successfully derive a set of hypotheses about the problems and needs of the user, but could not decide reliably on the required support. In consequence, an adaptive system has to be modest, and not fire off one-option solutions. It should "place the user in an information space" (Fischer, Henninger, & Nakakoji 1992) as system-initiated delivery mechanisms of help systems do. The users have to play an active role: They must make the final decision about adaptations, they must be able to select the class of adaptations (menu entries and/or key shortcuts), and they must be able to define the names of new entries, and so on. Our tests show that users are ready to process a list of suggestions only if they are relevant for their tasks and needs. The presentation of such a list increases the user's motivation to consider, and his awareness of, appropriate adaptations. Further experiments are needed to test whether, given a sample of adaptation items, the user is directed to go beyond the range pre-specified by the system and look for earlier adaptations or other adaptation features. Adaptation suggestions can foster an explorative use of the tools for customizing the application. In our test sessions, there is support for the hypothesis that, having a degree of freedom when confronted with adaptation tips, users prefer being autonomous and competent when configuring the system to being faced with a single proposal that can be accepted or rejected. This also facilitates our understanding of users who have experienced some adaptation proposals from the system and their developing expectations of further adaptation capabilities that are of a higher level of difficulty and are, therefore, not included in the system.

6.4. Performance and Control Support of Adaptation

6.4.1. Supporting the Adaptation

To perform an adaptation efficiently and effectively, as well as to visualize and eventually modify its results in later sessions, the user needs support from the system. Tests showed that such support should include at least the following features:

- It has to contain a short and clear introduction to the rationale of the adaptation, for instance, by tutorial.
- It has to present a comprehensive adaptation environment with selection and definition opportunities for adaptations.
- It has to offer an overview of completed adaptations: new or changed functions, new or changed system dynamics, new or changed system presentation.
- It has to allow for subsequent modification of adaptations.

We learned from observing the users that making adaptations is an iterative process. Sometimes, adaptations are triggered by a specific adaptation tip or initiated autonomously due to a specific task condition. Sometimes, they are modified, extended, or completed in the course of a quasi-excursion away from the current task, in which the focus of attention is customization of the work environment.

We partially identified the requirements as a starting point for the design of Flexcel I described early, in an earlier study about the use of adaptable systems (see Karger & Oppermann 1991; Oppermann & Simm, this volume). We improved on these initial concepts in the evaluation cycles and tried to support them with corresponding features. In the first version of Flexcel, we presented all information in one dialogue box describing what the user could adapt, why he might want to, and with what effect. This was insufficient because the user had no direct view of the modifications of the system. The modified function replaced the original function in the menu, which could then only be executed with its former effect by a key combination to be memorized by the user in the course of the initial adaptation. Definitions of new key shortcuts had to be entered without an overview of the already existing definitions. Use of a key that had already been defined resulted in a message beginning a question-and-answer dialogue. An overview of new key entries was only accessible via a separate menu option that none of the users actually used in the test sessions.

In Flexcel II, the design of appropriate adaptation features was better accomplished.[9] When prompting the user with an adaptation tip, the dialogue box shows the relevant parameters for the function to be adapted and a selection list of keys available for defining new shortcuts. Furthermore, it shows a free entry field for shortcuts and a text field for new menu entries. The user can either implement the proposed adaptation or employ the adaptation facilities to the given function (see Fig. 3.2). This adaptation environment is consistent for all possible access forms: from the adaptation tip, by clicking the "process" button; from the tool bar, by clicking the "adapt" button and selecting the desired function; and from the dialogue box of the function by clicking the control field "with adaptations" (see Figs. 3.4 and 3.5). The control facilities in the adaptation environment offer all relevant options and expose all relevant information to the user at a glance. Already defined adaptations are displayed in a selection list showing the corresponding parameter(s) of the function when selected. They are also integrated into the presentation of the system by being displayed in the menu below the original function entry. This was selected from a variety of options because it has the advantage of preserving the task-oriented context. The user only has to memorize the location of the basic function; all adaptations derived from this function can be displayed within this context and form a comprehensive family.

6.4.2. Giving Adaptations a Name

The test users understood the adaptation environment easily and were able to use the tools shortly after scanning the system. The only serious problem for the users was deciding whether to just define a key shortcut or also to add the new adapted functions into the menu, and how to name new menu entries. The adaptations appeared to be ad hoc modifications with relevance only for the task sequence at hand.[10] The naming of the new command was sometimes reduced to a quickly entered nonsense set of letters. After experiencing difficulties in remembering the meaning of these arbitrary menu entries, users criticized their own product, replaced the given names they had given, and started using better ones for subsequent adaptations, but even after this experience, the users had difficulties in finding really meaningful names. Only in very few cases could an underlying principle be recognized from the resulting names.

---

[9] The difference was due to the experience with Flexcel I, and to the more powerful facilities of the new Version 3.0 of EXCEL, which arrived as we were moving from Flexcel I to Flexcel II.

[10] This is possibly due to the test condition in the lab, and it would be a question for a long-term experiment whether the effect disappears when users define adaptations for their authentic tasks.

6.4.3. Criticizing Adaptations

Criticizing the naming of new functions can only be done by the user and is motivated by difficulties with unintuitive names. The current system's critique module neither proposes appropriate names nor criticizes user-defined ones, but there are other opportunities where an advisory component could help: for instance, in explaining the principles of composing useful names, such as, "Use characteristic and discriminating elements for entry names".

Other adaptation aspects might be criticized by the system interactively deducing breaks in the user's usual interaction pattern. For example, the system could evaluate whether a user identified as a "menu user" defines new functions with menu entries and a user identified as a "key user" defines new functions with key shortcuts. In the critique component of Flexcel II, constructive advice is prepared for seven possible improvements in using the adaptation facilities: "clear the adaptations", "use adaptations", "adapt on your own", and so on. In the test sessions, curiosity drove the users to take a look at the critique. This corresponds well with the design principle of the critique system (i.e., only to analyze longer periods of interaction and not to help with specific problems). To benefit from the critique, the user must use the system in longer term sessions than just, for instance, using a sequence of functions with identical parameter values.

6.4.4. Control Over the Adaptation Profile

When a system is adapted, the result is specific to a task and to a user, but this does not mean that the adaptations are only useful for that task or even only for that user. Adaptations may meet the requirements of a class of tasks for a group of users. Flexcel II has two supplementary entries in the file menu: "Save profile" and "Load profile". The profile stores all adaptations of the given system made by the user: all additional menu options, shortcuts, and parameter defaults. The user gives the profile a name that is displayed in the header of the main menu (to the right of the last entry). The user is always informed about the current profile and can load another one when turning to a different task. The profile is stored on an external floppy disk. To load a profile the user has to insert the disk and select "Load profile" from the adaptation menu. The profile concept separates the adapted system from the application and from the user's documents. It also gives the user more control over the privacy of sensitive data of the usage record that could conceivably be used to draw conclusions about working style, preferences, fuzziness, errors, and so on. The user can simply put the floppy disk with his profile into his pocket, if necessary.[11] The test users were unanimous in their appreciation of this feature because it provides both, comfortable access to user- and task-specific profiles and privacy. Further research is needed for the conceptualization and long-term evaluation of individual and group-oriented exchange facilities for adaptations and their acceptance. For instance, we are discussing the ideas of a "purse" for the exchange of profiles with successful adaptations within groups of users with similar tasks.

6.5. Discussion

The adaptive system presented here differs from others in that it provides mechanisms for adapting the user interface of a commercial system other than at the level of context-sensitive help (see Fox, Grunst, & Quast, this volume; Fischer, Lemke, & Schwab 1985). The spectrum of adaptations was identified and implemented based on empirical analyses of the potential of the application for adapting the interface to task- and user-specific needs. Adaptation features were implemented in an adaptive and an adaptable version to analyze the relative benefit of the two alternatives. The most important result of the study is that adaptive and adaptable systems are not alternatives. Adaptable features are not a solution, because—at least at present—they are not a common part of user interfaces, so users are not (yet) familiar with their existence, operation, and benefits. On the other hand, adaptive features are not a solution either, because they keep the users dependent on suggestions with respect to time and content of (tips for) changes. Users refuse to be at a system's mercy. This holds true for two dimensions. First, the suggestions made by an adaptive system can differ from the cognitive and operative expectations the user develops or has developed about the system. Second, the dependence on adaptivity produces emotional tension.

Such cognitive and operative interference has been demonstrated in other studies of adaptive systems. Mitchell and Shneiderman (1989) showed the disadvantages of dynamic versus static menus when the positions of menu entries were rearranged according to usage frequencies of specific users. The most critical point of studies like this, using quantitative performance criteria for the evaluation of adaptive features, is the time dimension (see Browne, Totterdell, & Norman 1990):

- When do adaptations take place: when a user begins to learn a system? when a user first begins to use the system for authentic tasks? when a

---

[11] This is certainly not an ideal solution for practical considerations because users have no pocket at all or no "free space" in their pockets. But even if the user has the floppy disk at his personal workplace, the profile is not accessible for centralized inspection and evaluation. Other technical solutions include password protection of the user protocols.

user already has real experience with the system and a system has data about the user?

- It is also important to determine the time for the evaluation of the adaptation effects: when the adaptation is being defined? when the adaptation is just finished? when the user has worked with the adapted system for a certain amount of time?

In our study we didn't use quantitative performance measures, but rather analyzed the user's cognitive understanding of and interaction with the adaptive features. In our experiments, the introduction of adaptivity follows after the basic understanding of the application, and the evaluation is concerned with the process of learning and assimilation. We found the adaptive and adaptable concepts to be most promising when the two features are designed to cooperate. The existing evaluations of adaptive systems test the adaptive effects more or less exclusively against static, rather than adaptable, systems (see Browne, Totterdell, & Norman 1990). We found the adaptive component to be best suited for preparing the user to adapt the system, for inspiring the user to reflect about the suitability of the application and for presenting clues as to when to turn from the domain task to the metatask of adaptation.

The emotional tension mentioned earlier induces the critical motivation for a latent attention to a transfer from adaptations by the system to the user's own active adaptation opportunities. The clues for adaptive suggestions (tips) should find a compromise between a massive interruption of the user's workstream with active help and merely mute potential with passive help (for a similar conception for design critic messages, see Fischer et al. 1992; Fischer, Henninger, & Nakakoji 1992). This compromise should ensure a "complete" utilization of the adaptive features' potential, while at the same time allowing the users to remain in control of their own working tools and working styles. The augmentation of the adaptation capability of the user is supported in the presented Flexcel II in three ways:

- It actively shows adaptation possibilities by adaptation suggestions. Thereby, it demonstrates to the user cases in which adaptations may simplify the work and so inspires the user to develop adaptations of the system in addition to those suggested by the system.
- It actively presents tips that propose that the user adapt independently or consult a tutorial he is not familiar with the adaptation tools.
- A critique component can be invoked by the user. The user's interaction with the adaptation tools is then analyzed, possibly resulting in suggestions as to how to use the tools more effectively.

One problem of the extension of an application with adaptation facilities was that some users, already having experienced the adaptive capability, tended to expect more from the system than it could offer. For instance, "When the system is able to recognize that I need a shortcut for this function, why doesn't it recognize that I need complete macros consisting of several function steps?" Macro learning could very well have been applied in the Flexcel context. For the user, this would be a small step; for the system developer, it would be a rather large one, from the actual implementation point of view. This is a question of the completeness and exhaustiveness of assistant capabilities. The implementation of features has to be complete, but it is subject to explicit reflection on what completeness means—for the user and for the designer.

6.6. Conclusion

There is no further development being done on Flexcel. We consider the research opportunities of Excel to be quite well exploited. One interesting extension, however, would be to combine Flexcel with a plan-recognition module, for instance, to enable the recognition of complete macros. However, too many hardware and software constraints make it difficult to proceed any further. The biggest problem is response time: The system is too slow due to the awkward (but only possible) way of communicating between Excel and the LISP software.

The ideas described in this chapter could easily be transferred to other menu-based systems. The knowledge-base and the presentation of suggestions are system independent and can be used in other systems, as well.

References

Browne, D., Totterdell, P., & Norman, M. (1990):
Adaptive user interfaces. London: Academic Press.

Brüning, I., Krogsæter M., & Thomas, C.G. (1991):
Realisierung von Adaptivitätsleistungen in einer kommerziellen Anwendung (Tech. Rep. No. 595). Sankt Augustin: GMD.

Carroll, J.M., & Campbell, R.L. (1986):
Softening up hard science: Reply to Newell and Card. Yorktown Heights, New York: IBM Watson Research Center, User Interface Institute.

Fähnrich, K.-P., & Ziegler, J. (1987):
Software-Ergonomie: Stand und Entwicklung. In: K.-P. Fähnrich (Ed.), *Software-Ergonomie*. München: Oldenbourg-Verlag, pp. 9–28.

Fischer, G., Grudin, J., Lemke, A., McCall, R., Ostwald, J., Reeves, B., & Shipman, F. (1992):
Supporting indirect, collaborative design with integrated knowledge-based design environments. *Human–Computer Interaction, 7*, 281–314.

Fischer, G., Henninger, S., & Nakakoji, K. (1992):
DART: Integrating information access and delivery mechansisms. Unpublished manuscript. Boulder, CO: University of Colorado, Department of Computer Science.

Fischer, G., Lemke, A., & Schwab, T. (1985):
Knowledge-based help systems: Human factors in computing systems. In: *Proceedings of the CHI '85 Conference*. New York: ACM Press, pp. 161–167.

Karger, C., & Oppermann, R. (1991):
Empirische Nutzungsuntersuchung adaptierbarer Schnittstelleneigenschaften. In: D. Ackermann & E. Ulich (Eds.), *Software-Ergonomie 91: Benutzerorientierte Software-Entwicklung*. Stuttgart: Teubner, pp. 272–280.

Mitchell, J., & Shneiderman, B. (1989):
Dynamic versus static menus: An exploratory comparison. *SIGCHI Bulletin, 20*, 33–37.

Newell, A., & Card, S.K. (1985):
The prospects for psychological science in human–computer interaction. In: T.P. Moran (Ed.), *Human–computer interaction (Vol. 1)*. Hillsdale, NJ: Lawrence Erlbaum Associates, pp. 209-242.

Oppermann, R. (1990, September):
Experiences with evaluation methods for human–computer Interaction.
Paper presented at the Fifth European Conference on Cognitive
Ergonomics (ECCE-5), Urbino, Italy.

Oppermann, R. (1991):
*Evaluation von adaptierbaren und adaptiven Leistungen im
Tabellenkalkulationsprogramm EXCEL* (Tech. Rep. No. 596). Sankt
Augustin: GMD.

Oppermann, R. (in press):
Adaptively supported adaptability. *International Journal of Human–
Computer Studies.*

Reiterer, H., & Oppermann, R. (1993):
Evaluation of user interfaces: EVADIS II – A comprehensive evaluation
approach. *Behaviour & Information Technology, 12*, (3), 137–148.

Thomas, C.G., & Krogsæter, M. (1993):
An adaptive environment for the user interface of Excel. In: W.D.
Graye, W.E. Hefley, & D. Murray (Eds.), *Proceedings of the 1993
International Workshop on Intelligent User Interfaces.* New York:
ACM Press, pp. 123–130.

Chapter 4
HyPLAN : A Context-Sensitive Hypermedia Help System

Thorsten Fox, Gernoth Grunst, and Klaus-Jürgen Quast

1 . Introduction

In this chapter, we describe the development of a help system, HyPLAN, for users of the spreadsheet Excel™ on the Macintosh™ computer. The design of this system, which integrates adaptive and hypermedia techniques, was driven by empirically identified user demand. By describing the development of the system, we hope to explain the interrelation between the analytical methods that we applied and the design of the various technical modules.

The developments were embedded in the research project SAGA, which deals with the design and evaluation of adaptive systems. Within the field of reasonable applications of adaptive concepts, context-sensitive help seemed to be especially useful. We therefore decided to develop an adaptive help system. This decision required that we select a target application. In order to provide context-sensitive help, we needed to be able to record and evaluate usage sequences of the target computer program. Unfortunately, applications available on the market usually do not come with access to their source code. On the other hand, preliminary tests of a prototype application—a logic based spreadsheet program (Spenke & Beilken 1988)—for which we could get access to the source code, showed that users had too much trouble using a coarse research system for their normal work. The interface of this prototype had not yet been ergonomically optimized through multiple redesign cycles. We were able to solve this dilemma when we recognized the possibility of obtaining records of Excel actions by using the application in its macro mode. Besides its ergonomically optimized interface, the level of cognitive demands that Excel makes on the user made the system suitable for the intended developments. Especially for casual users, Excel is neither too trivial, such that no problems can be expected, nor too sophisticated, such that only specialists can make use of the application.

In our research project, we decided not to amend unintelligible interfaces or functions by adding adaptive support concepts. Rather, we felt it would be more to the point to add help for difficult multistep tasks to applications like Excel, which represent the actual ergonomic state of the art. We see improvements as co-evolutionary processes yielding on the one hand smoother interfaces for elementary operations, and on the other intuitive help for complex operations that are not self-evident.

In order to identify a limitable spectrum of typical and significant work done with Excel, we selected *knowledge workers* (see Gottschall, Mickler & Neubert 1985), scientists, and managers, as target users. Field interviews showed a typical task for this target group is to extract significant data from journals and similar overview documents. Because such workers cannot foresee all important correlations and aggregations of information before actually scanning the data, it would often be inappropriate for them to delegate these tasks to spreadsheet specialists, who are unable to judge such relevancies; managers or scientists want to do these tasks themselves. They therefore need to be able to scan spreadsheet tables without being forced to waste valuable time (re)learning pertinent system functions. The target group could be specified as casual users needing complex functionality, such as compositions of commands, to extract specified items. This type of user is likely to forget previously learned application concepts. Therefore, knowledge workers seemed to be a group who would be especially interested in continuous on-line support for their task spectrum.

In interactive test sessions with pairs of Excel experts and target users, we tried to identify this spectrum of typical support demand. Although the project started with a focus on adaptive support concepts, these tests soon showed the importance of adequate presentation modes of relevant help. We could see, for example, that demonstration-based information was more readily accepted by the users than mere verbal explanations. Furthermore, evaluations of a first precursor of the help system—implemented in HyperCard™ and mainly consisting of textual help cards—showed that written text confused or strained test users facing real problems. Both findings hinted at presentation forms that integrate graphical and symbolic information and led us to design a help system that applies hypermedia techniques (HyTASK) before we started to develop an adaptive component.

Subsequent evaluations of this help environment clearly indicated that users also needed help in accessing the help system. Although applicable and easily understandable help was available, test users often could not get at it because they could not choose appropriate help items from the help system's long index list. We therefore put adaptivity into concrete form as a system module (PLANET) that selects actually relevant help items, based on evaluations of elementary Excel operations. The system we developed implements situation- and task-sensitive adaptivity by inferring the user's probable intentions and problems from his usage records. The inference results are exploited to help the user find relevant explanations.

In its final form, the entire HyPLAN system consists of two modules forming an adaptive hypermedia help environment for Excel:

- HyTASK (Hypermedial TASK support), a hypermedia tutorial and help system. This includes a network of hypertexts and animations explaining

Excel tasks that we found to be characteristic of knowledge workers. Help is given both visually and verbally through text, graphics, interactive animations, and spoken comments. The usefulness of these different communicative means had been seen in our preparatory studies of human— human tutorial interactions. We identified suitable scopes for technical applications of sound, text, graphics, and demonstrations, and evaluated these findings in subsequent versions of HyTASK.

• PLANET (PLAn recognition through activated task NETs), a knowledge-based plan-recognition system that evaluates records of Excel operations. PLANET identifies Excel work steps and infers the user's goals on various task levels. It then uses the results of this inference process to produce a context-sensitive ranked selection of help items.

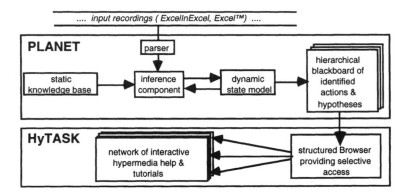

Figure 4.1:   The architecture of the adaptive hypermedia support system HyPLAN.[1]

This configuration enables HyPLAN to respond to nonspecific help requests with focused support (see Fig. 4.1).

The spectrum of plans that HyPLAN can recognize is, of course, limited to tasks formally represented in the knowledge-base of PLANET. Correspondingly, the help that the system can select and offer to the user is restricted to the set of tutorials and help topics available in HyTASK. The

---

[1]   ExcelinExcel is a duplication of Excel applying its macro facilities. It offers an interface identical to the short menu version of Excel supplemented by the "Table..." function. For the user, there is no deviation from the normal Excel. ExcelinExcel generates a protocol of usage sequences that is used by PLANET.

usefulness of the entire system therefore depends on the domain-specific relevance of the set of plans considered in both modules. Thus, empirical detection and modeling of notorious Excel use problems was a basic prerequisite of the developments. We empirically identified typical barriers to understanding, as well as efficient support patterns of human tutors. The latter findings were taken as models of successful help responses and guided the design of functionally equivalent technical concepts.

2. The Analytic Methodology Guiding the Design Process

In our research approach, we tried to integrate analytical methods and practical goals of system design. The goal of understanding reasons for (mis)understandings of certain computer concepts required analyses of tasks and related human cognitive processes. This necessarily involved reconstruction of the (possibly erroneous) conceptions guiding users' intentional behavior. Case-based discovery of Excel usage profiles was more important than the observance of standards for operationalized psychological or sociological tests.

The methodological approach that we applied therefore must be differentiated from common psychological and sociological methods, which focus, more or less, on the verification of theoretical concepts:

The studies were not based on laboratory experiments in the sense of psychological test design. Data influencing behavior are too complex to be analyzed according to statistical inference procedures. The contents of behavioral analyses cannot be restricted to cognitive processes and types of knowledge explicitly modeled in the cognitive sciences. For many relevant details of complex intentional interactions, there are no psychological concepts, models, or methods explaining or even predicting behavior and thought processes in real-world situations.

For example, inferences in realistic environments are highly context dependent. Specifying such contexts, however, requires metacognitive abilities to identify and apply the criteria relevant to the situation. Such metacognitive evaluations do not follow fixed rules applicable in every context. They depend, rather, on an ability to recognize important relations between objects and (inter)acting persons and to match the evaluations with one's own intentions. To a certain degree, these relations are dynamically changed by the actions themselves. The general ability to follow context changes and determine what is actually relevant is natural human behavior. Even if interactions are recorded on video, humans analyzing these interaction sequences are able to understand the mental processes behind the behavior. In this sense, human scientists (*coders*) can recognize and reconstruct detailed intentions, problems, and related evaluations by analyzing videotaped interactions between test users and experts. If they are familiar with typical aims of test users in the task domain, coders can relate verbal and nonverbal behavior details to these mental activities.

Essential to these procedures are in-depth interpretations. In contrast to everyday interpretations, scientific approaches require justifications of assumptions. To avoid subjective misconjectures, methodological measures have to be applied so that (other) coders can recognize false and insufficient interpretations. Discourse analysis (van Dijk 1985) provides methodological and theoretical models aiding in-depth interpretations. The main operational means for achieving objectivity is to split the interpretation into small, repeatable, and criticizable steps. A team of coders following discourse-analytic procedures has to demonstrate and criticize the adequacy of their descriptions by relating them to observable details. Elements of this procedure are transcripts and classifications of behavior units. We applied these discourse-analytical principles and procedures in the empirical studies through which we prepared and evaluated the design of HyPLAN.

In order to get some insight into troublesome Excel features, we chose test tasks that we hoped would produce natural problem-solving situations. The users' behavior was audiovisually recorded with as little interference as possible. We then selected parts of the interactions and analyzed them, applying discourse-analytic methods. In addition to more theoretical aims, such as identifying the dynamics of emerging and changing mental models, we had practical goals, such as identifying counterintuitive and complex Excel features. We applied user-relevant concepts of the task domain and Excel to determine thematic contents of the recorded behavior. To classify the interactions between tutorial experts and test users, we could relied on models of didactic patterns for (in)efficient help and critique (Jussen, Grunst, Dorn, Prinz & Kaul 1985, Rehbein 1980). Moreover, the psychological concept of *constructive interactions* (Miyake 1982) could be applied to look for and classify mutual criticisms indicating situations that were not entirely understood by (one of) the interaction partners.

The methodology applied in the studies of support concepts for Excel not only specified the analysis process but also covered the procedure of preparing and recording the test situations.

2.1. Preparing the Recording Sessions

We started our empirical studies by defining our particular analytic questions. In discussions with designers, software users, and psychologists, we planned test tasks that could be expected to disclose specific Excel usage problems for our target users. We modeled the test tasks by applying *task-analytic decomposition* (Phillips, Bashinski, Ammerman & Fligg 1988). Then we selected groups of target users and determined their Excel expertise on functions required or convenient for solving the tasks. We then made the test task concrete by matching it to data in the test users' actual work domain.

Combined task- and user-competence analyses allowed us to derive detailed hypotheses about difficulties users might be expected to encounter during problem solving. We thus anticipated task-relevant cognitive effects before the experiments took place. These expectation patterns guided the analyses of the recorded sessions. As a side effect they also influenced the configuration of the recording cameras. For example, if we could expect discussions about computer screen details the camera positions were set up so that gestures such as pointing to specific areas of a spreadsheet could be seen in the videotapes. When we had to focus on tutorial or constructive interactions between experts and test users the user cameras were positioned to capture gestures and facial expressions of the interacting partners.

2.2. Audiovisual Recordings of Task Processing

We tried to keep important behavioral details in their chronological and logical order. To this end, test users and the computer screen display were videotaped on a pair of synchronized recorders. A technical provision to avoid disturbances of natural interactions was to locate the recording staff in a separate room, where they could follow the interaction on video monitors. The two cameras were left stationary. Because there was no cameraman, the camera viewpoint had to be preset. The settings tried to show the entire scene, including the computer and the users. On the other hand, we tried to keep visual details of the interactions, such as facial expressions and gestures. We had to achieve good compromises between these restrictions and the requirements identified in the preceding task analyses.

There is good reason to be critical of videotapes as basic research data. The recording situation certainly influenced normal behavior, even though the cameraman was not present. Furthermore, the interactions were influenced by the test task and the selection of partners. How can this be brought into line with the authenticity criterion that seems to be a prerequisite for the discovery of mental activities? As already described, one provision for authenticity was that the contents of the test tasks be related to the users' actual work. Moreover, the level of cognitive demand was carefully selected so that the tasks were seen as neither trivial playing nor as too complex or irrelevant to the user. In order to guarantee natural personal relations between experts and target users, the test sessions were embedded in Excel courses. In these tutorials, the test partners became acquainted with each other and with training and helping interactions.

2.3. Analytic Reconstruction of Interaction Processes

Some of the crucial usage features of Excel were immediately evident from the tests, but we often saw problems in which the actual reasons for the users' difficulties remained unclear. We worked with the video recordings in order to try to identify the hidden aspects of these misunderstandings. The procedure started with attempts to obtain an overall picture of the session by applying the following three steps:

- By scanning and discussing the videos (two synchronized recordings of the test persons and the computer screen), the coders tried to identify the general gist of a session.
- Coders wrote a rough outline of the course of action with reference to video time tags.
- Coders selected sequences with implications for the study and marked them for further analysis.

In the more in-depth analyses of the selected interaction sequences, we applied discourse-analytic methods in order to model mental processes that might explain the observed behavior. The procedure consisted of three different treatments:

- Scoring of transcripts (see Rehbein 1980) of the verbal, nonverbal, and physical behavior of the interacting persons with respect to the actual computer displays.
- A chronological segmentation and classification of discernible interaction elements according to their purposes and contents.
- A comprehensive interpretation of the analyzed sequences evaluating whether they might be instances of more general interaction or thought patterns.

By applying these discourse-analytic procedures to the recorded sequences, we were able to cut the interpretation process down into small, explicit, and criticizable steps. Forms and categories that were used for transcripts and classifications oriented the coders and let them identify the reasons for their mostly intuitive judgments. Scanning of transcribed and classified sequences by other coders often produced insights into contradictions and inadequacies of the analytic reconstructions. This usually also changed the coder's general interpretation of the overall sequence of actions.

In order to classify recorded work steps as efficient or complicated, we applied the task-analytic descriptions of related task elements that were made before the tests began (see Section 2.1). Thus, we could categorize solutions chosen by test users to solve the test task as either *suboptimal* or *smooth*.

Some of the categories that we used to classify interactional intentions were taken from discourse analyses of pedagogical interactions (Jussen et al. 1985 Rehbein 1980). These studies apply a set of illocutionary functions classifying the intentions of interactional contributions. We classified not only the purpose of a behavior unit, but also the type of background knowledge applied by the acting person. Moreover, we tried to determine the type of background knowledge that was called up in the interaction partner in order to elicit an understanding of the problem at hand. We applied about 30 knowledge categories in order to classify used or addressed types of background knowledge. The categories and their coding specifications were continuously extended and verified, but only simple and rather irrelevant contents could be operationalized in general. Other relations needed contextual understanding and descriptions. The methodological procedures, thus, mainly had the effect of supporting and guiding the discursive interpretations of human coders.

Usually, groups of two or three coders were involved in the analyses of the video recordings. They mutually criticized their classifications and interpretations. They were urged to stay open-minded for relevant details not covered by classification categories. The dynamics of mutual critique and growing insight during interaction analyses is vividly described in Suchman and Trigg (1991).

In general, we needed to apply a method that provides strategies for verifying (necessarily) intuitive interpretations. In this sense, the reconstruction of mental activities relies on justified selections of analyzed scenes taken as representative. Descriptions and explanations have to be related to behavioral data. One scientific paradigm—among others—reflecting this epistemological requirement is *objective hermeneutics* (Oevermann, Allert, Konau, & Krambeck 1979). This research concept is based on three principles:

- Do not presume objectivity where, in fact, you have to interpret situations selectively and have to determine the relevance of certain data.
- Be aware of the requirement to inductively discover and name unknown details and processes, even though the perception itself—at least to a certain degree—depends on applicable conceptual categories.
- To provide explicit and thereby criticizable frames of perception and significance, relate the interpretative reconstruction as much as possible to pertinent theoretical concepts.

Thus objective hermeneutics does not deny the subjectivity of interpretations. Rather, intuitive insights are made open to criticism by relating them to observable data. For the descriptions, theoretical concepts modeling social and cognitive reality merely provide a fragmentary frame of reference.

3. The Identification of Working Concepts for the Support System, HyTASK

Once we had identified typical Excel problem areas using the techniques described, we developed the hypermedia help system HyTASK in design—evaluation—redesign cycles, beginning with analyses of human tutorial strategies. In these empirical studies, we tried to identify efficient human patterns for correcting, criticizing, and helping with crucial Excel features. We scrutinized successful ways of conveying to users how their goals could be matched to the spreadsheet as potential models of technical support.

3.1. Empirical Identification and Modeling of Adequate Support Effects

The design cycle started with investigations on how authentic tasks involving Excel were processed by the target users. In preparatory discussions identifying the users' expertise with Excel , some of the target users remembered that they had once learned certain advanced operations valuable for their work, but as casual users of Excel, they had seldom used these operations and had therefore forgotten them. In order to give them a basic understanding of crucial functions appropriate for the test tasks, they received introductory tutorials. Then test tasks were planned, that needed some of the advanced functions evidently relevant for the users' normal work. Because these tutorials were not directly connected to the test tasks, the users applied the information inefficiently and fragmentarily in the subsequent tests. We recorded the tutorial interactions with a human tutor and analyzed them for potential models of appropriate help. After this preparatory phase, the users had to solve similar tasks by applying the newly learned or relearned Excel concepts. In these situations, the accompanying expert changed from a guiding tutor structuring the interactions into a specialist helping with difficulties as they occurred.[2] Figure 4.2 shows the experimental setting of these studies.

---

[2]   As a means of knowledge acquisition, these (constructive) interactions are an advanced variant of more traditional forms of protocol studies relying on think-aloud utterances of a single user (cf. O'Malley, Draper & Riley 1985). In think-aloud situations, the test person is usually missing a natural purpose for the utterances. Two people dealing with a complex matter quite naturally communicate hypotheses, questions, answers, critiques, and justifications, which more accurately reveal their mental models.

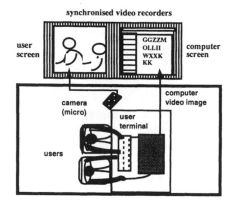

Figure 4.2:   A test design disclosing use problems and pertinent human help.

The mental inferences could be related to four different types of observable behavior:

- Working on the task via computer input (test persons).
- Indicating trouble (test persons).
- Suggesting pending support or critique (expert).
- Giving hints (expert).

All types of behavior included physical actions, and verbal and nonverbal elements.[3] Interpretations of these observable details allowed us to reconstruct the underlying cognitive processes. We examined the experts' actions in terms of both substance and form, as a model of behavior potentially convertible into technical equivalents. We evaluated the users' reactions as important indicators of the effectiveness of the experts' help suggestions.

3.1.1. How Users and Experts Focused on Problem Areas

One of our most clear-cut findings was that nonverbal activities were of greater importance than we initially imagined. Interactions between the experts and the test users about actual problems usually turned out to be multistep processes initiated by nonverbal behavior. A user working on task elements and no longer quite sure of how to proceed started to indicate loss of control by changing his or her posture and the pace of keyboard input. The expert was

---

3   Nonverbal behavior includes facial expressions and gestures, whereas physical actions mainly include the computer-related operations.

alerted and actively started to narrow possible gaps in understanding. Similarly, if the human expert identified potential problems by matching the user's known intentions against his course of actions, he also started to indicate nonverbally that support or a clarification dialogue should be initiated. The expert held back comment until the user indicated a willingness to be interrupted. These activities rely on general and hardly understood interactional abilities to evaluate context changes with respect to the goals that are actually relevant in a given situation. We rated these human cognitive capabilities as unsuitable models for guiding the design of a technical support system. A knowledge-base able to simulate this complex behavior was judged to be an unrealistic goal. Instead, we tried to select appropriate details of efficient strategies for mutual orientation, and to transform these behavioral elements into workable technical concepts.

With this as our goal, we could uncover typical ways that users expressed their problems with certain Excel concepts. The simplest means applied by users to convey to the expert the scope of problem was to point to graphical interface details. For example, one test user accidentally touched the "Split window" button and noticed the change in the mouse cursor's shape. He evaluated this as something meaningful but unknown and initiated a pertinent clarification dialogue. Through quite indefinite questions combined with deictic mouse cursor gestures, the user was able to inform the tutor very precisely about his actual difficulty. Thus, a basic communicative barrier was neutralized. The user could focus on unknown system features even though he was missing applicable terms for the pertinent Excel features. Users in real trouble are most unlikely to be able to identify gaps in understanding by imagining and naming probable causes.

The most difficult problem that the experts encountered proved to be distinguishing real mistakes from new but reasonable ways of performing a task. Discussions with experts indicated that they often used analogous episodes as frames of reference. Test users with other experiences sometimes had to convince the expert that they had no trouble performing a certain type of task, and that the strategy they were following was not suboptimal.

3.1.2. How Recognized Problems Were Cleared Up and Solved by Experts and Users

When problems had been mutually recognized and the test user had consented to be informed about certain Excel features, the tutor often cleared up the problem through demonstrations of example solutions. The expert accompanied his demonstration of task steps with verbal comments illustrating their relevance to the user's goals.

After giving a graphical demonstration of how to solve a task, the expert sometimes tried to communicate a deeper understanding to the user of how the spreadsheet program worked in the particular case. Instead of demonstrations, now verbal explanations became prominent. The tutorial experts most often described the system metaphorically. Typically, Excel was depicted as someone taking in information, transforming the contents into desired results, and presenting these results at the interface. Crucial functional features of Excel, such as differences in copying cells with relative, absolute, or mixed links to other cells sometimes were depicted by graphical metaphors, such as "moving rubber-band connections" versus "rigid pointers". The users especially picked up on these types of metaphorical expressions. When they faced similar tasks, they remembered, talked about, and followed the ideas condensed into these descriptions. The studies thus showed an efficient two-step pattern:

- Build up a basic and intuitive understanding of new concepts by actually demonstrating how to apply them in order to achieve certain goals.
- Encourage generalizations of newly learned concepts by explaining the system's functioning in graphic, metaphoric terms.

These findings guided the design of a functionally equivalent help system.

3.2. Technical Transformations of Findings and Evaluations of the Derived Design

In our transformations, the demonstration-based teaching concept became an interactive hypermedia tutorial system. In a manner analogous to the identified human explanations, HyTASK conveys information mainly through interactive hypermedia presentations showing how to solve certain types of tasks. HyTASK presents Excel "scenes" in order to build up an immediate understanding of error-prone details. A presentation of task steps accompanied by spoken verbal comments about reasons and pitfalls introduces efficient ways to solve certain tasks. This turns out to be an efficient analogue to the human didactic practices identified.

If information is transmitted entirely through demonstrations, however, learners show worse transfer effects than with verbal explanations (Palmiter & Elkerton 1991). Therefore, we tried to integrate the experts' metaphorical descriptions of system operation, which—according to our preliminary studies—supported the generalization and active use of the learned concepts. As a technical transformation, the "how to do it" demonstrations (short help demos and longer tutorials) were supplemented by "how it works" demonstrations of crucial Excel features. These graphically animated and metaphorically rich illustrations explain crucial Excel features. Evaluations of system prototypes confirmed that this promotes active use of the Excel concepts conveyed by HyTASK.

Both types of demonstrations present the information units in typical work contexts. The system provides specific task contexts as film-like sequences of work progress in Excel. In some of the Excel sheets illustrating the work steps, HyTASK presents graphically emphasized buttons marking potential trouble spots. The user can click on these buttons to get more information about the corresponding system functions. This is one of the ways in which HyTASK offers the user context-specific help access reflecting task states and foreseeable help demand.

Follow-up tests on successive versions of HyTASK uncovered more and more subtle faults. In order to identify fine-grained cognitive adjustments of the support system, we first had to correct coarse problems. For example, we first had to modify the presentation technique so that relevant information could be recognized. Then intelligibility could be evaluated and modified. Moreover, generally counterintuitive features had to be fixed before specific demonstrations of intricate details of work could be taken into account. So, we first had to identify and put into practice working strategies for building help explanations, such as "start with a description of goals which can be achieved by using the introduced functions". We could then fine-tune the explanations by integrating task-specific forms of demonstrations.

In order to avoid redundant treatment of topics, the HyTASK development aimed at modular information units that could be used in several task contexts. It turned out, however, that many troublesome features of spreadsheet applications like Excel are related to complex combinations of specific details. Explanations then have to take into account the contextually relevant aspects. To form a useful set of elementary and complex help items avoiding redundancy and confusion for the user, we first had to identify clusters of information relevant for specific task contexts. Then we checked the value of this information in analogous problem situations. In order to achieve a working modularization, we tried to break down explanations into smaller units that could be combined to capture more complex problems. This could not be achieved in general. Some Excel usage problems appear only in specific complex tasks. In its final design, the support environment therefore includes a mixture of very specific high-level demonstrations of notoriously intricate tasks and short explanations of simpler concepts applicable to several types of spreadsheet calculations.

The linked interactive demos offer "scenes" including bundles of information mediated through sound, graphics, texts, and animated presentations. In designing the help environment, we could combine these cues so that they could induce a spectrum of suppositions relevant for different users in different contexts. Because the demonstrations have the character of scenes, users with specific interests can easily ignore irrelevant aspects and avoid becoming confused. The amount and many-sidedness of information that can be conveyed to different users this way cannot be achieved by (single-channel) explanations via text and graphics. Users exploited the information associatively in often unforeseen ways. In order to prevent them from getting lost in an information maze, the scenes constantly have to give clues enabling users to evaluate states when they are viewing a film. HyTASK, in this sense, mainly induces such evaluations through spoken comments, and provides intuitive controls (temporary buttons) allowing users to switch between information units.

According to our evaluations, HyTASK was transformed into a smooth system after about 1 year of test and modification cycles involving four graver and a lot of minor changes. The system's animated help and tutorial films were approved by the users. Minor criticisms still referred to control problems concerning unintuitive backtracking links after having visited related help. Four main flaws of the hypermedia system were identified and fixed during the test redesign cycles:

(a) One of the early and most clear-cut results of the studies was that written text should be avoided if possible. In preliminary tests we recognized that the text-based precursor of HyTASK was inappropriate. In addition, the first graphics-based version of the help environment, implemented in HyperCard, had extra textual help that popped up when the users touched certain "hot spots". This seemed to confuse the users. The textual information mode apparently interfered with the users' processing of the otherwise graphical information. On the other hand, users readily accepted even complex explanations and successfully matched them with their own intentions or problems if they were presented in the form of demonstrations of the necessary work steps.

- So, after the first redesign cycle, short textual help cards were designed and supplied with buttons leading to animated demonstrations. HyTASK now uses written text mainly in the form of short descriptions on these cards informing the user of which topics are dealt with in the selection of help films. In even later versions of HyTASK, integrating the screen recording software MediaTracks™ as a development tool facilitated the production of animations decisively. The execution of a task could, moreover, be shown more realistically. Even subsemantic levels of work, such as movements of the mouse, are now displayed. The user sees how every state changes into the next and can therefore remember the execution of task elements as a (perceptual) motor program (Wright 1990) corresponding to mouse movements. The visual presentations are supplemented by spoken comments about reasons for and pitfalls of work steps, including difficult Excel functions.

(b) A further shortcoming of the first HyperCard version, prior to the integration of MediaTracks, was its confusing dynamics and control concept. In this version of HyTASK the animated demonstrations were actually a series of Excel screenshots that could be presented in either a step-by-step or a self-running mode. Different entry points of users addressing the support system required different modes of presentation. Because we could not foresee wether a user wanted to see a presentation of a whole task or just specific details, the user had to decide on the best mode of presentation. Although the users usually had an intuitive idea of what they wanted to see, they were often unable to understand the request to specify how a demonstration should be presented.

-   After we had replaced the HyperCard screenshot sequences by MediaTracks recordings of the Excel operations, we offered only continuously running films. These animations are controlled through a simple control panel resembling that of a video cassette recorder. It intuitively allows stopping, backing up, and forwarding a demonstration. Furthermore, the speed of the presentation can easily be adjusted. The users quickly became accustomed to these command facilities.

(c) On some cards, the HyperCard version of HyTASK presented touch- and click-sensitive icons and areas leading to further information. These "buttons" were visually marked as framed graphical elements that were highlighted when touched with the mouse cursor. Because the demonstrations presented a lot of unknown information, the users frequently overlooked the specific function of these graphical buttons, and pertinent hints were missed.

-   In the MediaTracks redesign, we therefore implemented the contextual access to further information through temporary text buttons, previously implemented as icons. The demonstration pauses if such "doors" leading to other demonstrations are offered. The user can decide if he should continue to watch the current explanation or switch to the suggested side films by clicking the button. Later, as the HyTASK information space became more and more complex, this "doorway" concept turned out to be a useful idea. The decomposition into films linked by conspicuous doors whose relevance could be intuitively assessed seemed to be a good way to help orient the users. The shortcomings described so far could be fixed by modifying the hypermedia presentation and interaction concepts.

(d) The most serious limitation, however, was related to the fact that the system often did not sufficiently support finding the appropriate help topic. This finally required that we develop a further module complementing the hypermedia system. In all the versions of HyTASK, the user could enter the help environment through a click-sensitive index list of tutorial and help topics. The most crucial HyTASK usage problem was connected to this initial access mode. Without advice from a supervisor, the system was used as little and ineffectively as written handbooks applying the same entry concept. Users indicated that searching for possible help through long index lists was seen as a diversion from the scope of their duties, and was therefore refused.

- We therefore decided to develop a plan-recognition system, PLANET, that offers the user a narrowed down, contextually relevant choice of help topics. The system identifies the user's probable intentions by evaluating his elementary Excel operations and maintains a picture of the actual help demand by adaptively taking into account changes in the work in progress.

4. The Development of HyTASK

4.1. The System Architecture of HyTASK

The final version of HyTASK, comprises three connected HyperCard card stacks —Helptopics, Navigator, Browser— and a collection of interactive films (see Fig. 4.3).[4] The individual help cards of the stack Helptopics and the interactive films serve as the actual information nodes and are linked together in a network. The Navigator and the Browser consist of a single card each and serve as access and overview windows. The Navigator constitutes the original access method via index list. The Browser provides a more flexible method for scanning the help topics, and incorporates the context-sensitive selection of help topics provided by the plan-recognizer, PLANET.

4.1.2. The Browser

The pivotal function of the Browser is to provide the primary access level to the whole support environment. The Browser appears after the user hits the "Help" key, and its content (selective access to help topics) is configured adaptively, that is, it is context sensitive. It consists of three columns of help topic names. The center column contains a list of, at most, five help topics selected by the plan-recognition module. Each topic is a "hot spot" (i.e., access to the corresponding help topic card is provided by clicking on it). Furthermore, by merely touching a topic in the center column, the user gets more information about its meaning and embedding in corresponding task contexts: A short task-related description of the topic appears in a field below, related subtopics pop up in the lefthand column, and help items representing more complex task units involving the highlighted topic appear in the righthand column. This arrangement of help topics according to task levels is de-

---

4   For the presentation in this volume, the original German interface has been translated whenever useful and necessary for understanding.

termined by a directed graph ("topic graph", see Section 4.1.6), which covers all available help topics. Figure 4.4 shows a sample Browser configuration with three center topics and the subtopics and supertopics associated with the highlighted topic "Formulas".

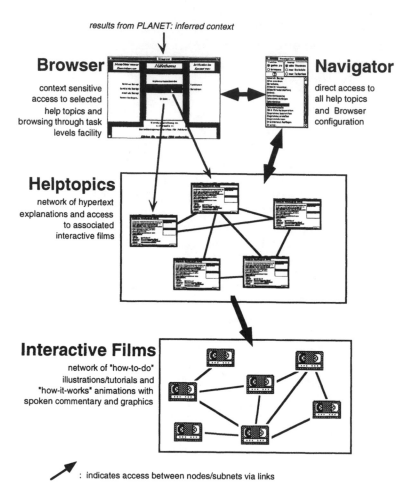

Figure 4.3:   Connections within the HyTASK architecture.

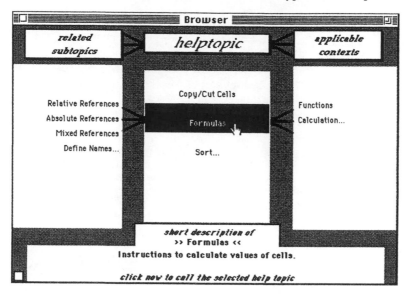

Figure 4.4:   A Browser configuration.

The Browser also allows the user to shift the side columns into the center by clicking on the pertinent column. This feature allows browsing of the topic graph by following the links determined there. In the new configuration, the user can once again directly access topics or browse further. If there are no related subtopics or applicable contexts for a topic, the respective columns just remain empty, and no more browsing in that direction is possible.

The Browser presents two different types of contextual embedding. First, as selected by the plan-recognition module, PLANET, the list of items related to the user's actual task context is put into the center column. Second, each center topic is embedded in a general task context according to the topic graph. Both kinds of embedding provide cues for the user to judge for himself "do I know this?" or "do I need this?".

4.1.3. Helptopics

The main Helptopics stack consists of 45 help cards. These cards correspond to the actual help topics: each contains a general description of the topic and access to its associated films. Each card contains an extended topic name as a header and a short textual explanation (see Fig. 4.5). In the first all-text version of HyTASK, this text was more extensive and was seen as the primary means of explanation. Now, it merely serves as a means of orientation for the user to evaluate the relevance of the offered films to his goals. Various buttons on the card provide basic interface functions. They enable the user to return to Excel, to track backward within the help environment, to search for strings within Helptopics cards, and to activate a metahelp mode for the use of HyTASK itself, that is, objects become touch sensitive and a short help text appears as long as the user moves the mouse across them.

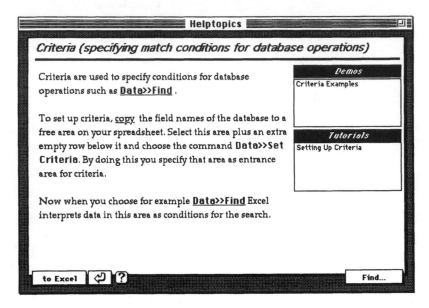

Figure 4.5:   A help item of Helptopics.

The description illustrates the topic by way of a typical example. Terms corresponding to Excel features, such as names of menus or dialogue buttons, are made conspicuous by using the system font Chicago.

All help cards are linked. The underlined phrases in the textual explanations represent click-sensitive areas leading to related help topics.

Two click-sensitive lists headed "Demos" and "Tutorials" represent links to the interactive films showing how to perform Excel task elements related to the topic of this card. Depending on the complexity of a topic, these demonstrations are either short examples (i.e., demos) or longer introductions (i.e., tutorials).

During our design cycles, as existing parts of HyPLAN were refined, the number of items within Helptopics grew. We started with some specific help items for our experimental tasks. Now, Helptopics consists of 45 cards covering the whole functionality of our test environment, "ExcelinExcel". Thus it covers all the commands of the short menus settings of the original Excel, including the commands for handling databases and the command "Table...".

4.1.4. Interactive Demos and Tutorials

Both kinds of interactive films are basically screen "recordings" of Excel tasks (i.e., spreadsheet calculations). They are accessible via a corresponding help card within Helptopics. There may be more than one film in either category on the card, and some films appear on several cards.

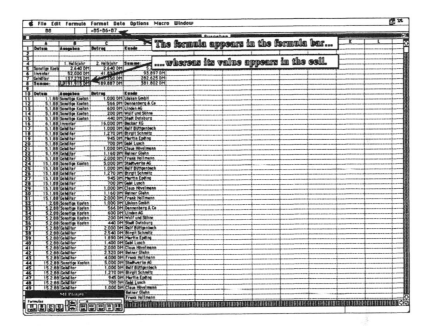

Figure 4.6:   Screenshot of a demo with graphical elements and the control panel.

The films start with a verbal description of typical goals achievable by applying the concept or command that is about to be demonstrated. They then introduce a topic by presenting screen recordings of Excel tasks, accompanied by spoken comments in a "how-to-do" manner. Graphical elements and textual cues appear in order to highlight important aspects. Figure 4.6 shows a screenshot of a demo, including arrows and text pointing out details. The films may contain links to related films (if any) accessible via buttons that temporarily appear in the presentation. As the number of help topics grew, so did the number of interactive films. Eventually, 59 demos and 8 tutorial films were produced, integrated into the network of already existing films and connected to the corresponding topic cards.

As the number of help topics grew, so did the number of interactive films. Eventually, 59 demos and 8 tutorial films were produced, integrated into the network of already existing films and connected to the corresponding topic cards.

The user can control the presentation of a film in the help environment by using the MTPlayer™ control panel shown in Fig. 4.7. It is comparable to a video recorder and enables the user to pause, continue, and stop the film; proceed stepwise; move fast forward and backward; rewind the presentation; and control the speed of the presentation. Furthermore, the panel can be hidden and allows backtracking through the films[5] already seen.

Figure 4.7:   The MTPlayer for controlling interactive films.

Demos and tutorials differ as to their length and elaborateness. Each help topic is exemplified by at least one demo not exceeding 25 seconds and accompanied by one or two spoken comments. Only a few demos contain links to other films. Usually, they are closely related to the text on the topic card and relatively independent of other demos. They are intended to provide on-line offers for briefly picking up, as well as refreshing, system concepts needed for the actual task. The user's attention should remain focused on the real work.

On the other hand, the basic function of tutorial films is to convey more complex system features. The user's attention should switch from solving a

---

5   Backtracking within the facilities of the MTPlayer is possible, but very unintuitive and error prone. The user has little feedback about the currently running film and no information about films already visited. Rebuilding a steering unit and integrating a history component would have improved the overall usability of the system, but would have been tedious and distracting from focal developments.

task to learning new items. Tutorial films take at least 5 minutes, forcing the user to mentally leave his actual work context. In contrast to the demos, they contain many links, both to other tutorials and to demo films on associated topics. Naturally, the time taken for tutorial excursions increases if the user studies these linked films. For example, details elaborating on a topic are put into additional films if the amount of information would interfere with a first intuitive reception. Access to these associated films is offered temporarily at appropriate times in the basic-level presentations. Links with a different rationales appear at the beginning and the end of the tutorials. When starting a tutorial, the user gets a hint about related subtopics that are prerequisites for understanding the current help topic, and is offered access to corresponding preparatory films (see Fig. 4.8). He might take a look at these films first, after identifying probable gaps in his knowledge. Similar links are displayed at the end of tutorials. Here, the user can access more complex topics that include the demonstrated topic as a subtopic. He might take these links if he decides to learn more.

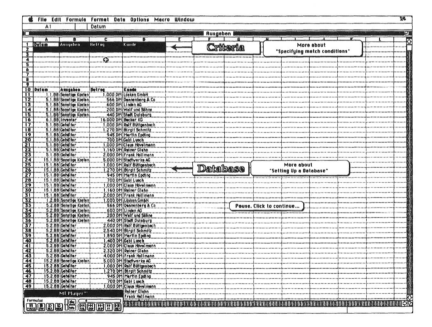

Figure 4.8:   **Tutorial screen offering links to preparatory topics.**

Both of these latter types of links correspond to a graphical structure that describes the relationship between all help topics on different task layers. (See Section 4.1.6 for a more detailed description of the "topic graph")

The presentation form of all links within films is a button with arrows pointing to the potentially unknown interface elements or screen contents. On the buttons, short phrases, such as "More about: Setting Up a Database", are used to sketch target items. Whenever buttons show up, pauses are provided so that the user has time to decide whether to access the extra films. This technique of making links visible turned out to be a more obvious invitation to take a look at additional demonstrations than were mere graphical highlights of the regions, because the latter did not indicate their extraordinary function within the presentation. Also, single words naming the buttons rarely enabled the user to decide on the relevance of the expected information.

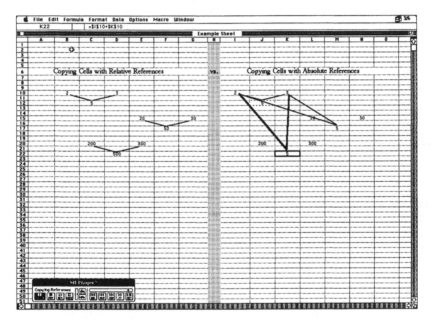

Figure 4.9:   Screenshot of an animation visualizing the "moving rubber band" metaphor.

Within the HyTASK demos, most of the messages are transformed into "how-to-do-it" films, as explained earlier. In addition to conveying this "how-to" expertise, some of the demos are in the form of "how-it-works" films, which illustrate the internal functioning of the system through animation.

This was a transformation of the tutors' "how-it-works" explanations in our experiments. For example, the complex system operations associated with the command "Table..." are depicted by moving graphical elements and related remarks. The user sees how Excel handles input information and what he should accordingly expect. Figure 4.9 shows a screenshot of such an animation, where a "moving rubber band" is animated in order to explain different consequences of copying cells with relative or absolute references. Within the animation, absolute references behave as stretching rubber bands when the source cell is copied, whereas relative references are shown as rigid pointers.

4.1.5. The Navigator

Within the screen presentation of the entire help environment, the Navigator window (see Fig. 4.10) is located on the right-hand side of the Helptopics or the Browser window and can always be activated by mouse click. It serves as a kind of landmark in the system by giving an overview and providing access to all available help topics in the environment. It consists of several alphabetically ordered indexed topic lists. The user can choose the complete list of help

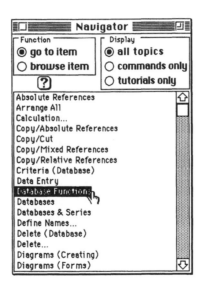

Figure 4.10:   The Navigator window functioning as a landmark.

topics, the list of help topics referring to commands, or the list of tutorials. Having selected an item in the list, the user can directly enter the associated topic card (go to item) or enter the Browser and configure it with this item (browse item). As with Helptopics, a meta helpmode is available, explaining the use of the Navigator.

4.1.6. The Topic Graph

The *topic graph* is a structure reflecting the hierarchy of all help topics according to different task levels. It serves mainly as a basis for controlling the three-level presentation of the Browser, but also for the links appearing at the beginning and the end of the tutorials. In this graph, a directed link is set up from one topic to another, if understanding the first is a prerequisite for understanding the second. For example, to understand the command "Calculation...", a user of Excel has to know the "Formulas" concept. Moreover, the links in the graph can also represent conditional relations between topics determining temporal interdependencies. For example, the arrows from "Set

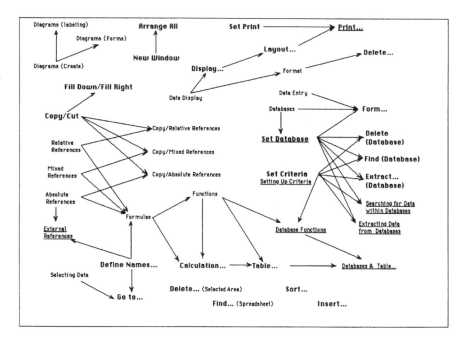

Figure 4.11:   The topic graph.

Database" and "Set Criteria" to "Find" reflect that the user must perform these earlier operations in order to enable the later ones. The complete topic graph is shown in Fig. 4.11 Note that not all areas are reachable from one another.

Items written in (the bigger) Chicago font correspond to Excel commands. The (smaller) Geneva font indicates items that are related to specifically problematic combinations of elementary features. The underlined topics are more complex tutorials. Also note how the Browser configuration shown in Fig. 4.4 corresponds to the links of the topic "Formulas". The left-hand and right-hand columns in the Browser consist of topics, having an arrow, respectively, to and from the topic "Formulas" in the graph. Browsing the graph can be described as moving a virtual window along the links.

The user never gets an overview of the complete (and confusing) topic graph, but only of contextual collections of items representing coherent segments of the graph. Links in the graph reflect, rather, different types of relevant relations between help items in order to let the user induce contextually appropriate insights and inferences. Providing locally consistent information instead of illustrating general relations between help topics is the intended purpose of the graph.

4.2. How HyTASK Can Be Used

Example 1. Suppose a novice uses Excel in order to sort a (just created) table according to the contents of its first row. He might be unfamiliar with the "Sort..." command, and therefore uses "Cut" and "Paste" to reorganize his table. After frequently repeating these two operations, he might get the impression that an advanced system like Excel probably provides a more efficient solution for such a task. Aware that help is available by striking the "Help" key , he addresses the support environment. PLANET is able to infer the user's actual intention (sorting his spreadsheet) by evaluating the use of multiple "Cut" and "Paste" operations and checking constraints confirming that the target areas of the operations imply a sort operation.[6] The center column of the Browser configured by PLANET therefore contains the help topics "Cut/Copy" and "Sort...". The user immediately matches the offer with his goal and obviously selects "Sort...". Having entered the corresponding help card, he might briefly reassure himself that this is actually the correct topic. Knowing the film concept or assuming that this might be interesting, he starts the film belonging to the card. The film shows him how to select a target area, and tells him that he must choose the command "Sort..." in the menu

---

6   Obviously, not every usage of sequential copy and paste commands is a sort operation.

"Data". Then, the example shows the central process of informing Excel about desired organization criteria by feeding parameters into the pop-up dialogue box by clicking appropriate cells on the sheet. Having seen a table being sorted as a result of these actions, the user returns to his own table and applies the learned sequence of steps.

Example 2. This example is taken from our evaluation studies. An intermediate user has set up a database. In order to look up specific data, he then chooses the command "Find…". This causes an error message requiring that he specify criteria. Having seen "Set Criteria" in the "Data" menu, he simply chooses this command. This causes a second error message informing the user that the area for search criteria is not correct. Now he addresses the help environment. The Browser window, which then appears, contains the topics "Set Database" (which the user has already done) and "Find…". He moves the cursor over the topic "Find…", and the side columns fill with related help items. To the left of "Find…", the subtopic "Set Criteria" appears. He shifts the left columns into the center of the Browser and selects "Set Criteria" to enter the related help card. The card briefly describes how to define and select the area for criteria. He realizes that he did not copy any field names, returns to his work, and corrects his mistake.

Example 3. A casual user familiar with the support environment may want to use the command "Table…", but does not remember exactly how to arrange the data appropriately. He will probably strike the "Help" key , ignore the Browser, and directly enter the Navigator window. There he activates its "Commands only" list, selects the command "Table…", and immediately starts the demo film accessible from the help card. The film first shows how to select an area before calling the command "Table…". He pauses the film and mentally matches his own spreadsheet with the example table. Having recognized how to arrange the data, he leaves the help environment without running through the rest of the film.

4.3. Design of HyTASK

4.3.1. The Choice of Tools for Prototyping

The demand-driven design–evaluation–redesign cycle requires a thorough selection of software tools. The typically large number of design changes require a development system to support smooth (re)implementation of desirable features (see Gould 1988). In our case, these features were aspects concerning the presentation and structuring of information.

Evolutionary redesign also requires that implementations be kept as modular and intuitive as possible. Modifiability should be favored over features such as processing time or disk space. Laborious implementations that have to be changed or are rejected after evaluation cause a lot of frustration or, even worse, block necessary changes because of economical considerations. Every version of the system should be seen as a development platform for the next generation and be open to extensions.[7]

One more criterion influencing the choice of design tools is that they must reflect the cognitive mode of information relevant in the target domain. This means that appropriate data types, such as text and graphics, can be integrated into the system, as well as dynamic presentations (e.g., animation, sound, and video). Our studies of efficient human support showed spoken comments and animated demonstrations of a task to be indispensable information forms, as were written text and graphics. On the other hand, video was considered an obsolete type of data for information presentation within HyTASK.[8]

Regarding our target machine, the Apple Macintosh, we found that Hyper Card was an appropriate high-level tool that meets the adopted criteria. Using HyperCard, moreover, guarantees to a certain extent that basic state-of-the-art interface features are provided. HyperCard is a kind of object-oriented development and run-time system that handles documents called *stacks*. These stacks consist of cards on which information (graphics, text, sounds) and interface elements (buttons) are placed. The functioning of the interface elements (such as playing a sound or changing the presentation of a card) can be specified in associated scripts using its programming language, HyperTalk™ (see Winkler & Kamins 1990). For example, with a script "jumping" from card to card containing graphics, cartoon-like animations could be created. Connections between cards, also called *links* can be designed with different techniques, such as buttons and touch- or click-sensitive lists. The presentations of cards, demonstrations, or meta-information (help mode) differ in terms of their possible interactions with the user. Every type of presented information requires specific control facilities. Scripts written in HyperTalk support a wide spectrum of presentation and interaction concepts. Furthermore, they are easy to read and therefore easy to change, rendering possible short design–evaluation–redesign cycles.

---

[7] This holds, especially, for help and training systems for domain-specific tasks that have to be maintained.

[8] If visual details of real-world situations are essential in order to impart crucial concepts to the user, the system must support video. However, within our target situation, the scenes to be visualized— namely, the processing of spreadsheet calculations—could be restricted to the computer itself.

In the first version of our system, we presented illustrations of tasks performed in Excel as sequences of cards showing crucial screenshots. The evaluations showed that continuous presentation and user-controlled timing of these help illustrations are essential features. Both had to be fixed in our illustrations. By applying this newly derived criterion, we replaced this part of the whole system by MediaTracks "films". MediaTracks is an efficient tool for recording continuous screen sequences and running them as interactive presentations. To implement interactivity, the designer is also provided with tools to cut the film; add pauses and sound; and combine them with additional text, graphics, and links to other films. During presentations, a built-in video recorder-like control panel provides intuitive means for tuning the speed. Thus, presentation and control features of the help environment identified as desirable could be realized in an even smoother way than within Hypercard "cartoons". The MediaTracks films went well with the nonanimated part of the help environment, which was still implemented as Hyper Card stacks. Special HyperTalk commands (XCMDs) can be used to start the presentation of MediaTracks films and to return to HyperCard again.

For hidden communication between the Macintosh applications and the plan recognition system, we used QuicKeys™. This tool allows keystroke macros to be defined for sequences of actions, such as menu commands, mouse clicks, keystrokes, and window handling. We used it not only for communication facility, but also to build a "clean" interface between the applications. The setting enables test users performing their tasks in Excel to call the support system (and initiate hidden communication) via the Help key, which prevents them from feeling that they are leaving their work context. The general technical criteria for our system stated earlier—modularity, flexibility in design, and the integration of various data types—as well as the choice of tools led to the insight that our system would be a hypermedia system. Not only the initial requirements, but also the later evaluations, showed, that techniques of structuring and presenting data according to hypermedia principles were crucial for our design purpose. Note that we did not plan to design a hypermedia system from the beginning of our research.

4.3.2. HyTASK as a Hypermedia System

General Aspects of Hypermedia

The basic idea behind hypermedia is to distribute information into nodes and to connect them via links (see Nielsen 1990b). The concept of hypermedia encloses two important properties: First, these nodes can represent arbitrary data types; and second, they can be connected in a flexible and unrestricted manner.

Hypermedia documents can establish nonlinear forms of information presentation and therefore differ from normal text documents. The free connectivity between nodes distinguishes the hypermedia concept from database concepts.

Sometimes, authors distinguish between *hypermedia* and *hypertext*. Within hypertext documents, information is in the form of plain text and graphics, whereas hypermedia documents also use sound (speech and music), dynamic graphics (animation), and video sources. In this respect, the notion of hypermedia subsumes hypertext, but it is quite common to use the term *hypertext* to designate both (see Nielsen 1990b). However, we prefer the term *hypermedia* to emphasise the rich communicative means implemented in HyTASK.

While navigating hypermedia environments, users can switch between nodes or single information subunits in a flexible manner. Links may provide rich forms of interactivity in order to exploit the possibly heterogeneous information sources. They may appear in various forms, such as underlined words or phrases within textual presentations, named buttons, or icons. Furthermore, they can be implemented using various techniques, such as short pop-up texts or selective replacements of single parts instead of changing whole nodes. In a graphical interface, links are usually click sensitive: The user can move from one node to another by clicking it on. In addition, they can be touch sensitive, which means that some action is initiated by just moving the mouse across a node. Both kinds of visible elements on the screen are also called *hot spots*.

The pros and cons of the hypermedia concept are widely discussed. Hypermedia encourages alternative structuring principles for information (see Waterworth & Chignell 1989). It has the potential to combine even heterogeneous information into one medium in a more natural way for both representation and perception than do sequential media, such as text. Presenting information using the combination of multiple media is considered a prerequisite for faster information perception, faster and deeper understanding, and a more sophisticated use of long-term memory.

These advantages concerning the presentation of information go hand in hand with the advantages for the access to information that is structured and presented by hypermedia networks. The following characteristics are regarded as its major advantages compared to information access in linear media (see Duchastel 1990):

- Nonlinear information access enables the user to navigate through the network in a sequence that is not anticipated by the designer.
- Variable information access achieved by different link presentations and implementation styles (e.g., icons) has a greater appeal to the interest-driven exploration of the user.
- Integrated information access combines various information sources (e.g., text, pictures, sound) into one medium.

- Easy access to information units leads to a better effort-to-interest ratio during information retrieval.
- Free access achieved by the offer of multiple links supports user control and thus promotes motivation, which is especially important in computer-based instruction.

Addressing Problems of Hypermedia Usage Within HyTASK

There are also some pitfalls related to the use of hypermedia systems. In contrast to books or other linear media, there is the risk that the user will become disoriented while navigating in the information space. The user may not be sure where he is or where to go in the complex net of information: He is "lost in hyperspace"(see Nielsen 1990a). In order to avoid this and to help the user to keep track of his location, landmarks, such as nodes reachable from everywhere; overview diagrams; and a history component can be integrated.

- Within HyTASK, the Navigator, which is accessible from everywhere, serves as this kind of landmark. Furthermore, the overall structure of HyTASK is very simple. There are only two different types of nodes representing (a) a network of static information (Helptopics, see further on) and (b) a network of dynamic information (interactive films). The two networks are joined. Thus, users only need minimal processing concerning their position in the overall structure during navigation.

Particularly in hypermedia sytems serving as learning environments, the perception of information (i.e., the learning task) may suffer from a deficit in the cohesion of the structure (see Duchastel 1990). A cohesion deficit may appear when the user tries to integrate possibly heterogeneous information, presented through various media, into a coherent structure. During navigation of a hypermedia system, learning is achieved by subsequently integrating newly perceived information with previously perceived information and existing knowledge. A lack of minimal cohesion can cause the learner to fail to achieve the necessary integration and therefore miss the learning goal. On the other hand, incoherence between contiguous information units (i.e., the linked parts) requires extra processing on the part of the learner to bring the information together. A certain degree of such extra processing can work for the learner, because it is necessary for the deep understanding of a topic and elaborated learning.

- HyTASK addresses this problem by using consistent examples within its presentations and a linking that depends on expectable errors and the help contents.

Another possible drawback of hypermedia usage is the loss of the actual information demand. In contrast to the situation of getting lost in hyperspace, where questions, such as "Where am I?" or "Where can I go?" arise, the question posed here is "What did I actually want?". Users may lose track of their original goals, as a result of a search in the information space that is too long and extensive, even if this search is focused on relevant aspects. This is especially true for hypermedia systems serving as help or support environments. This situation (i.e., the loss of the conceptual point of view) was also detected in our analytical studies.

- One of the major aims of HyTASK has been to work on this problem. The system helps the user to find a help for his problem quickly by providing a focused entrance into the help environment. This first kind of adaptivity is achieved by the combination of plan-recognition and hypermedia access techniques. Furthermore, orientation effects within HyTASK are achieved by helping users with their tasks and with notorious problems within their tasks. This second kind of adaptivity is achieved by a design of the help contents that is closely related to tasks and notorious problems.

Demand-Driven Hypermedia Design

According to our concept, the process of navigating through a hypermedia information space to target information can be seen as entering rooms, scanning their doors, considering their probable goals, choosing the best direction, and confirming the decision after having entered an information room. The success of such support environments therefore depends on how well where-to-go decisions during navigation are supported. Door inscriptions have to indicate what can be expected behind them. For this purpose, the user's pursued aims and possible disorientations have to be considered within the design process. Therefore, design requirements for interfaces to hypermedia systems are highly user and task-specific. There is a lack of general standards and guidelines for tailoring hypermedia systems to particular needs (see Waterworth & Chignell 1990). Relevant differences concerning usage are hard to characterize and, hence, to match to design principles. Models for efficient design processes is an important research issue in hypermedia design. In our opinion, a design of hypermedia systems that maximally embodies the advantages (e.g., intuitive presentation and partitioning, and flexible access to complex information) and minimizes the possible drawbacks (e.g., getting lost or losing the information demand) during usage can only be achieved in an evolutionary design process that models and fine tunes user- and task-relevant aspects.

Our own approach in the development of the support environment, HyPLAN, relied on the already described studies of specific help requests and pertinent human support in the target domain. The detected tutorial help patterns explaining difficult multistep tasks were transferred stepwise into technical equivalents. So, studies of successful human behavior and the evaluation of prototypes guided design of the following main system aspects:

- Form of explanations: film-like illustrations and animations, instead of text.
- Spectrum of contents conveyed: combinations of topics and problematic aspects.
- Entrance into the help environment: narrowed access and intuitive circumscription of contents.
- Structure and linking of the overall system.

Dynamic Links as Hot Lists

A smooth technique for implementing the dynamic links of the Browser, as well as the Navigator's index list, is supported by HyperTalk. The respective lists of topic names are implemented as a text field. Its single lines contain the names of cards within the Helptopics stack. Each line acts as a *hot spot*: It is linked to the topic card of Helptopics named accordingly. The HyperTalk script of the card field responsible for this behavior is shown in Fig. 4.12. Using these *hot lists* and implementing links via card names upholds the modifiability of the system as new topics had to be added incrementally during redesign. The same technique is used to realize the links to demos and tutorials on the topic cards, and in the design of the Browser for the center column for presenting the inference results of the plan-recognition module.

```
on mouseUp
    put the value of the clickline into cardName    -- line
content
    if cardName is not empty then
        if the hilite of btn "browse item" then
            -- go to the Browser card...
            go to stack "Browser"
            -- ... and configure the center column with it:
            browse cardName
        else
            -- go to the respective help topic card
            go to card cardName of stack "Helptopics"
        end if
    end if
end mouseUp
```

Figure 4.12:   **A HyperTalk script of a card field serving as a hot list.**

5. The Identification of Working Concepts for the Plan Recognition System, PLANET

5.1. Analyzing Human Plan Recognition

Evaluations of HyTASK showed that the hypermedia presentations and intuitive explanations provided an acceptable form of help. The users were able to navigate through the help environment and to transfer the explained course of actions to their real task problems, once they had found the appropriate help topic, but in many situations, users had problems when they entered the help environment in finding the help topic via the index list. Matching vague actual problems with the broad spectrum of help topics seemed to be a cognitive load incompatible with the processing of real tasks. Quite normal intuitive mapping of problems to topic names seemed to be blocked by stress.

It was at this point that we decided to expand HyTASK to include adaptive, focused help. We therefore needed to find a suitable research basis for the discovery, development, and evaluation of adaptive help. We started with discovery experiments in order to find a model for desirable forms of user-sensitive support. In so-called "Wizard of Oz" tests (see Hill & Miller 1988), involving Excel and the already developed HyTASK, we simulated an intelligent help system by having a group of experts, hidden from the user, generate context-sensitive entries into the hypermedia system. Our goals were to determine how users react to various forms of help, as well as to determine the rationales used by experts when giving advice to typical users. In this way, we could evaluate help proposals and methods for recognizing users' intentions before implementing an appropriate module.

In our experiments, a pair of users (novices or low-level experts) together tried to solve authentic tasks slightly beyond their routine competence. Two human Excel experts watched the users' screen via a monitor terminal in an adjacent room (see Fig. 4.13). Both groups had concurrent control of the Excel application running, but the experts were only allowed to act in certain ways that were set in advance. If the users asked for nonspecific help (by pushing a "Help" button), the experts were instructed to offer a selection of help topics they guessed to be appropriate to the situation. Furthermore, if the experts watching the test persons identified a usage problem that they thought should be corrected immediately, they were allowed to interrupt the test subjects with specific help offers, textual dialogue, or both. The communication was in the form of a dialogue card (text area on screen) that appeared on demand. The experts did not know the actual test tasks nor the solutions. Nonverbal and verbal interchanges between the two users were hidden from them. They had to rely on observable computer operations to recognize the users' goals and problems. To provide help for the users, the experts could make up a list of help topics to be displayed on the card, which the users could click on to start the corresponding HyTASK help. During the experiments, the users worked normally with Excel. The actions and reactions of the experts appeared to them as an "intelligent assistant".

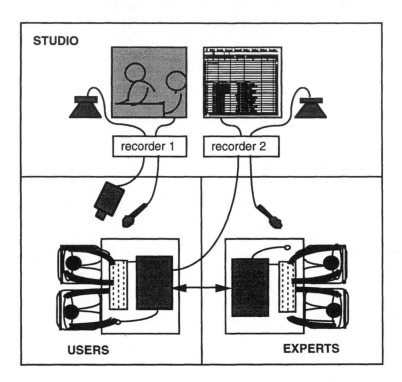

Figure 4.13: Set-up for the Wizard of Oz experiments.

On both sides of the "wall" related constructive interactions (see Miyake 1982) took place. The users communicated about the task and ideas for possible solutions and about the (re)actions of the "assistant". The experts tried to identify the users' goals and problems. They mutually criticized each other's assumptions and help strategies.

Each experiment was recorded on two synchronized video tapes. On the audio track of the first tape, the verbal communications of the users were recorded. The audio track of the second tape was used to record the verbal communication between the experts. The video track of the first tape recorded the computer screen as visible to both groups at all times. The video track of the second tape could be used to record the user group, or the expert group, or both groups using split-screen recording. The recordings could be and were replayed together. So, in analytic reconstructions, every judgment could be evaluated against observation of both groups.

5.2. Technical Transformations of Human Plan Recognition

We found that active help offers were rejected by the users surprisingly often. Especially when users were still following a (possibly false) idea about how to solve an actual task step, they were really annoyed by interruptions. On the other hand, when they were aware of having no working mental model, they readily accepted help suggestions. This finding made us aware of the problematic dynamics of active support and led us to give preference to context-sensitive but passive solutions.

Besides this restrictive finding concerning the desirability of active support, three different types of successful identification strategies applied by the experts could be determined. Two of these were reflected in the design of the plan-recognition module, PLANET:

(a) The experts made use of domain-specific knowledge to recognize the users' intentions.

They differentiated, for example, between names for people, departments, products, time-spans, and costs. Furthermore, they used domain-specific knowledge about the relevance of certain relations between these concepts. They used the structure and semantics of the identifiers in the colomn and row heads of the table to interpret the rationale of their use. For example, when the user defined a region to be used as a database, the experts used the meaning of the datafield identifiers to recognize an operation that extracts a dataset out of the database. A technical equivalent of these aspects of human expertise was judged to be an unrealistic goal. The great amount of application-specific knowledge used in a very specific way varies with changing situations. We could see no possibility of defining general patterns that could be applied to the syntactic information available from user action records in order to interpret the users' activities in depth.

(b) The experts used task action patterns to recognize the users' intentions.

Taking into account the whole spreadsheet, after only a few steps the experts were often quite sure—although sometimes mistakenly—of whether the users were having trouble, what goal they were pursuing, or both. They matched the users' actions against abstract patterns of complex operations. We identified patterns for good action sequences, as well as faulty and suboptimal sequences. To understand the action in depth, the experts again made use of domain-specific knowledge. We tried to incorporate this idea of abstract task patterns into PLANET's knowledge-base. The experiments helped us to identify how detailed a task pattern needed to be to be used for recognition. At the other extreme, the experiments showed that the experts used more abstract, complex chunks of actions. The way in which experts combine chunks of actions to fill more complex action patterns directly influenced PLANET's process of collecting information. We observed experts who collected these action chunks and then checked the compatibility of all subactions involved. An analogous bottom–up recheck mechanism can be found in PLANET's control strategy. On a higher level of meta-cognition, we had problems identifying the strategy the experts used to focus on and select plans for evaluation in the subsequent steps. The experiments allowed us to prove the existence of such a process but not to determine how this process works in detail.

(c)   The experts used a simple mechanism to generate hypotheses.

Experts used single significant functions to immediately infer complex user goals. For example, a call of the function "Extract"—although its use may have been illegal in the actual situation—allowed the experts to assume that the user wanted to handle a database and extract information from it. Furthermore, certain error messages caused by inadequate input indicated specific—sometimes extensive—gaps in users' knowledge. These "cue" indicators were picked up on by the experts to get an initial focus and to establish a reduced set of possible hypotheses. These types of cue subactions turned out to be rather reliable, and we integrated the idea into PLANET as an hypotheses generator that reflects the experts' observed strategies. The designer of PLANET's knowledge-base for a particular application can mark single subactions as such cue subactions.

Guided by these experimental results, we built PLANET's knowledge-base as an ascending network of elementary, intermediate, and top-level task nodes. The task domain is represented as a hierarchical net of related task nodes on seven levels. Each task node contains place holders for constituents. Task nodes at the lowest level of the net can be instantiated by recorded actions. When instantiating the patterns, the constituents have to take into account temporal order and associated constraints. In a spreading activation process, confirmed actions trigger all patterns that include the identified action units as elements. This ascending identification process spans all task levels. The results are recorded on seven corresponding blackboard layers. Besides completed actions, hypotheses of probable plans are identified and noted on another blackboard. Hypotheses are generated by an algorithm that follows marked links between subactions and actions. Which subaction–action relation is marked to be used for hypothesis generation can be determined in PLANET's static knowledge-base by the designer.

If help is requested, PLANET actualizes the evaluation of the recorded actions. It then uses a heuristic to select maximally five of the actual inference results (primarily high-level hypotheses) noted on the blackboards as help topics to be presented to the user.

6. The Development of PLANET

6.1. Introduction and Overview

In order to recognize the users' intentions, we had to represent observable action patterns in a form called *action plans*. For the recognition task, we applied a technique by which action plans are modeled as multilayered symbolic nets. The nets are used by an inference component as a signal conductive system to recognize the modeled action plans. Using spreading activation releases the knowledge-base designer from handling interrupts and overlaps, which can disturb a recognition process. Because the inference process can be stopped, continued, and backed up without completely resetting and recomputing the inference status, this technique can be used to model Undos. The inference component generates an action history, which stores recognized executions of actions and subactions on each level. This history will be evaluated by other components. We describe the use of PLANET within the Excel help system HyPLAN (see Quast 1993) to offer the user a selection of help topics sensitive to his or her task handling. The Evaluator analyzes the computed history in order to determine crucial usage problems and detect suboptimal task executions that occur during an application program.

PLANET is a first step toward a universally applicable method for recognizing action plans in various domains. It introduces a technique for modeling action plans with multilayered symbolic nets. Apart from modeling the plans' structures with nets, the plans are defined by sequencing the subactions and constraining the parameter values that occur at run-time. The inference component spreads activation impulses from the input nodes constituting user actions, to the output nodes, which constitute complex plans. In this way, the net serves as a signal conductive system for the recognition process. Recognition problems induced by parallel plan executions that interrupt each other can be ignored using this net-based approach. For each action and plan, the inference component stores the recognized executions in such a way that additional components can trace each execution back to the intermediate and elementary action executions to determine crucial usage problems and to detect suboptimal task executions. Because each plan execution is stored in a fixed set of data entries, the process can handle Undos of action executions by deleting the entries generated during the spreading of the action executions.

6.2. Plan Recognition: State of the Art

In this section, we outline some performance criteria for algorithms that recognize action plans, and discuss the PLANET system and other state-of-the-art recognizers with respect to these criteria.

6.2.1. General Problems of Plan Recognition

There are different types of syntactic interactions between action executions that the algorithms must take into consideration.

1a.) Interrupted executions: The execution of an action plan may be interrupted by the execution of another action plan and continued at an indeterminate later time (Fig. 4.14).

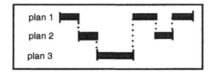

Figure 4.14:   Interrupted executions.

1b.) Branched executions: The execution of an action plan is interrupted by a second action plan and continued after the interrupting execution is finished (Fig. 4.15). This interaction is a special case of interrupted executions but more easily handled because interrupted executions can be stacked.

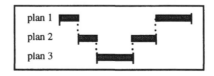

Figure 4.15:   Branched executions.

2a.) Overlapping executions: Two executions overlap if there exists a set of subactions that is part of the first execution, as well as part of the second (Fig. 4.16).

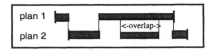

Figure 4.16:   Overlapping executions.

2b.) P/s-overlapping executions: A p/s-overlap of two executions exists if there is a sequence of subactions that is a suffix of one execution as well as a prefix of another. The subactions in the overlap are part of the first execution as well as part of the second (Fig. 4.17).

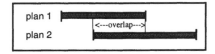

Figure 4.17:   P/s-overlapping executions.

PLANET handles all forms of interaction just listed. Because PLANET is an inference network, the activation of a node triggers the activation of all other nodes to which it is connected. In this way, actions can be interpreted as parts of several higher level plans, and, thus, overlapping actions do not interfere with the recognition process. The problem of interrupted executions is solved in PLANET by storing the activation states for each plan as the recognition proceeds. The state of a plan does not change when actions of other plans are executed and recognized. An "interrupted" execution remains in its state and will be continued if appropriate subactions follow.

6.2.2.   Other Plan Recognizers

REPLIX

In REPLIX (see Dengler, Gutmann & Hector 1987; Kempke 1986), action plans are described as sequences of actions. Each plan action consists of a command and some variables (e.g., cd dir: clear the directory named dir). In order to match the user's actions with an action plan, the actions must be executed in the same order as the commands occur in the plan. During the matching process, the variables of the plan actions take on the value given by the operands of the user action. The values of variables must be consistent for all subactions. The first command of a plan is the Start command. If the Start command has been matched, the recognizer tries to match subsequent user actions with the suffix of the plan. Once the Start command of a plan has been recognized, each of the subsequent user actions has to match the plan.

There can be several plans active in the recognition phase at one time. Because each action can be interpreted by all of these plans in parallel, the recognizer handles p/s-overlapping (see Fig. 4.17). To handle interrupted executions, each plan description holds a set of Interrupt commands. Interrupt commands are Start commands of other plans that are explicitly entitled to interrupt the execution of a particular action plan. An Interrupt command appearing in the sequence of user actions interrupts the process matching the plan (the plan changes to status "asleep"). If the interrupting plan is recognized completely, the matching of the interrupted plan will continue (the plan changes to status "active"). The recognizer handles branched executions, but not the more general case of interrupted executions (see Figs. 4.14 and 4.15). Because REPLIX can handle branched executions, it is able to handle overlappings more complicated than p/s-overlappings, but cannot handle overlappings in general.

If, during the match of a specific plan, an action does not fit in the plan and it is not an allowed interrupt action, the match fails. To prevent such failures, each plan description may contain a list of Ignore commands. Ignore commands do not affect the recognition of a plan. REPLIX input thus consists of a description of the plans given as a sequence of commands plus the set of interrupt and ignore commands.

MODIA

The recognizer MODIA (see Schwab 1988; 1989) enables the designer to define plans in a "programming" language called DELTA. *Terms* can be created that describe action sequences, alternatives to reach the goal, parallel executions, and repetitions of actions. Terms can be combined into more complex langistic expressions. MODIA transforms the given plan description into a realization tree. The structure of this tree corresponds to the building structure of the DELTA expression by which the plan is defined. MODIA matches the realization tree's leaves with the next expected user action. The values assigned to variables in a realization tree are validated for consistency.

The system is designed to support automatic continuation of incompletely executed plans. The recognizer handles p/s-overlapping executions. Branched and interrupted executions are handled explicitly by skip-lists (see Ignore commands in the previous section).

RPRS

The RETH Plan Recognition System, RPRS (see Miller 1990), is based on the RETH system (Allen & Miller 1989) and TEMPOS (Koomen 1989). TEMPOS provides tools for defining complex time sequences of subaction executions valid for plan schema. In RETH, the actions and basic plans are defined in hierarchies of abstraction using the mechanism of inheritance. There is a hierarchy of actions and a hierarchy of basic plans. Both range from the abstract to the specialized. The designer can describe actions as abstract patterns and refine the descriptions by adding conditions that constrain the subactions or by substituting new subactions. Koomen defined abstract basic plans with reference to abstract actions and specialized plans with reference to specialized actions.

Basic plans can be combined into more complex plans. TEMPOS will validate an attempted plan match by matching the unbound variables of the plans with the user's action. This is implemented by a backtracking procedure that searches for a consistent assignment of values to variables, taking into account the restrictions given by time constraints defined in TEMPOS.

The system manages p/s-overlapping executions. Similarly to PLANET, the recognition process builds plan instances that can be completed by actions occurring at any time. It therefore handles interrupted and branched executions.

CONPLEC

The CONPLEC system (see Berger 1990) is based on the parallel distribution of signals in a net of primitive units. The units of the net represent distinct information entities, and the connections between the units have a meaning defined explicitly by the knowledge-base designer. For this reason, CONPLEC is not a connectionist system with distributed representation, although it works in a similar manner. Elementary actions and composed sequences of actions are represented locally by single nodes. The inference component implements a bottom–up spreading cycle of activation. A plan is recognized if the node representing the plan exceeds an activation limit. Nets are structured in four layers. The first layer represents elementary commands, and the fourth layer represents action plans. The command nodes will activate the plan node only if the command nodes are activated in a predefined linear order. The two layers in between connect a plan node and some command nodes according to the action–subaction relationship. To enforce a given order of command, CONPLEC uses a technique based on an extension of connectionist networks with precondition and exception links (see Chun 1986). The links between two nodes can be weakened or strengthened by the activation state of a third node. It is possible to constrain a command node to activate a plan node only if all commands of lower order and no command of a higher order have been executed.

Because several different nets can be used for recognition at the same time, the system handles overlappings and branched executions. The designer may connect plan nodes to handle interrupted executions. The interrupted plan will keep its activation status while the interrupting plan is being executed.

6.3. The Components and the Flow of Information in HyPLAN

The user works with Excel by entering commands at the dialogue interface (see Fig. 4.18). In the background, Excel's macro facility records the user's actions in a file that is read by PLANET. We had to extend the Excel macro facility to get a more informative recording that includes error messages not normally recorded by Excel. Using the macro programming features, we built a simulation of Excel's interface called ExcelinExcel. The added interface layer enables the system to record all user actions and Excel reactions. ExcelinExcel and the normal Excel macro recorder generate a usage record that can be read by PLANET. PLANET's parser separates each entry of the resulting record into a command string and a parameter string. Together with a sequence number, determined by the occurrence of the entry in the sequence of all entries, the command and parameter strings form an elementary event. Such events contain the name of the action executed, the parameters of the action, and the sequence number. If the entry CHOOSE(Z1) occurs as the Kth entry in the recording, the parser will form the event (CHOOSE-CELL, Z1, K).

The elementary events are transmitted to the inference component, which attempts to match them with the action plans stored in the static knowledge-base. While matching events with plans, the inference component generates instances of plans. Each instance represents a partial or complete execution of a plan. A partially recognized execution is called an *hypotheses* if it contains cue subactions indicating a high probability that the user wants to execute this action plan. The instances are stored in the dynamic database.

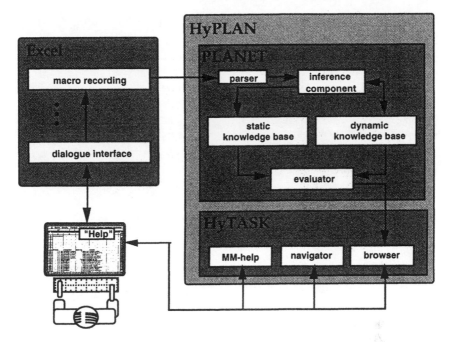

Figure 4.18:   Architecture of Excel/HyPLAN.

The results of the inference process can be extracted from the dynamic knowledge-base at any time. If the user hits the "Help" button, an evaluating component maps the set of completed plans and generated hypotheses to the set of available help topics in HyTASK. The selected topics fill the Browser (see Fig. 4.19)

In the middle column, the Browser offers the list of help topics determined by PLANET. The columns to the left and right show topics that have a more specific or a more general meaning for the topic active in the middle column. The user can scan the list, move topics from the left or right column to the middle column, and select the topic he or she needs. The subsequent dialogue is mainly determined by the way the information is presented in HyTASK (see Section 4 for details).

6.4. PLANET's Action Plan Model

In our use of PLANET, the static knowledge-base contains compiled descriptions of Excel action plans that have been identified by observing and analyzing user sessions with Excel in the Wizard-of-Oz experimental control setting (see Section 5.2).

These action plans identified in the real world can be described by structure, sequence and content.

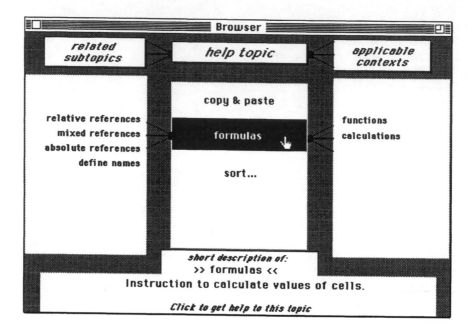

Figure 4.19: Browser of HyTASK.

6.4.1. Structural Descriptions of Action Plans

The structural description of an action plan determines how an action can be divided into subactions. For the designer of the knowledge-base, it is helpful to use operation trees to visualize how complex action plans can be synthesized out of elementary actions or, conversely, how complex action plans can be divided into elementary actions. The formalism of And/Or trees (see Nilsson 1980) can be applied to model the structure of a plan.

The nodes of a tree represent the goal of the plan, the Or-subtrees constitute the alternative ways to reach the goal. Each And-subtree represents one alternative made up of lower down, previously defined action plans (subactions). Finally, a complex action plan is reduced to a number of elementary actions (leaf nodes), which can be mapped directly to the possible events (user commands). The order of occurrence of the subactions within one alternative does not determine the order of execution of the subactions.

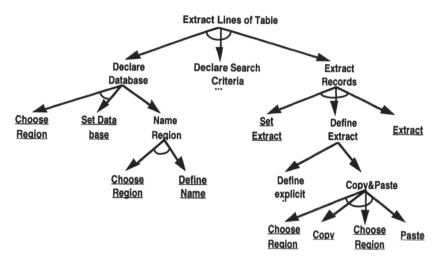

Figure 4.20:   Operation tree of "Extract Lines of Table".

For example, to extract lines from an Excel table, the user has to declare a database, declare search criteria, and make the actual extraction (see Fig. 4.20). A database can be declared by choosing a range of cells and executing the menu function Set Database. An alternative way to declare a database is to name a chosen region with the string database using the menu function Define Name.

Defined actions and plans may occur in several plans. The existence of these "universal" actions motivates the designer of the knowledge-base to describe the action space in a modular manner.

In the same way that single actions can be modeled by And/Or trees, the entire set of all action plans (union of all actions) can be modeled by an And/Or graph. Combining all the operation tree modules will result in an operation graph that consists of input nodes representing elementary actions, output nodes representing high level action plans, and intermediary nodes representing operations on a specific level of the resulting taxonomy. The elementary and intermediary nodes can be accessed by more than one high-level node.

6.4.2. Time-Oriented Descriptions of Action Plans

It is not sufficient to describe an action plan by structure alone because nothing has yet been said about the order in which the subactions must be executed. Accordingly, the description of alternatives has to be expanded to include the timing requirements; for example, here is a module of the operation tree from above:

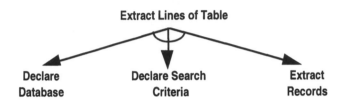

Figure 4.21:   Operation tree of "Extract Lines of Table".

The execution of the operations Declare Database and Declare Search Criteria in Fig. 4.21 are independent of time order, but both operations have to be executed before the operation Extract can begin. For convenience, we restrict:

time(Declare Database)   <   time(Extract)

time(Declare Search Criteria)   <   time(Extract)

We can distinguish between plan-oriented and interplan-oriented time restrictions. *Plan oriented* time restrictions limit the execution order of subactions occurring in one alternative of an action (see Fig. 4.21). To formulate restrictions, from this point of view the action can be imagined as a separate unit, and the timing interactions with other action plans can be ignored.

On the other hand, sometimes there are timing interactions of subactions belonging to different plans. Generally, *interplan-oriented* interactions occur during execution of several plans that overlap in their subactions.

Take a look, for example, at the plans Cut&Paste and Name Region, which can share common Choose Region subactions:

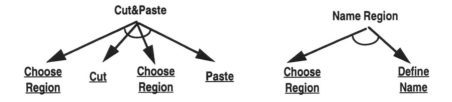

If the user executes a sequence of events (listed in the middle row of Fig. 4.21), a matching problem may occur.

Figure 4.22:   Interaction by overlap.

An uninformed simple matching algorithm will recognize one execution of the action plan Cut&Paste. Moreover, it will recognize two (!) executions of the action plan Name Region because Choose Region was executed twice, although the user really executed just one Name Region action (the first region is no longer active). To solve the problem, we must restrict the composition of Choose Region and Define Name to be legal only if no additional Choose Region execution occur between subaction executions. Such a constraint will rule out Name Region match1 (see Fig. 4.22).

6.4.3. Content-Oriented Descriptions of Action Plans

So far, the structural and time-oriented description of an action plan determines which subactions belong to it, constrains the order of the subactions, and takes care of timing with other plans. However, the elementary actions of an action plan are normally executed with parameters. The values of the parameters may become important in identifying an action plan. For instance, parameter values determine whether an action may be interpreted as a subaction of a plan.

For example, the operation tree in Fig. 4.23 takes into consideration that Excel allows declaration of a database by choosing a region and executing the procedure Define Name, which is a Name Region action. The action Name Region may be interpreted as a Declare Database action only if the Define Name action names the region with the string "database".

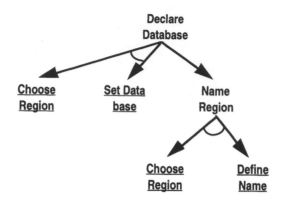

Figure 4.23:   Operation tree of "Declare Database".

Constraints based on the parameter values of an execution are called *semantic constraints*. We can differentiate between single-action–directed and multiple-action–directed semantic constraints. *Single-action–directed* semantic constraints determine a constraint condition for a single subaction in the context of an action plan (e.g., in Fig. 4.23). *Multiple-action–directed* semantic constraints are used to check whether the parameters of several subaction executions are compatible in the context of one action plan.

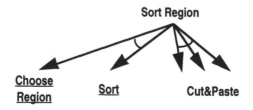

Figure 4.24:   Operation tree of "Sort Region".

For instance, a region can be sorted by using the menu function Sort. The other alternative for Sort Region is modeled by three Cut&Paste operations, as shown in Fig. 4.24.

This reflects the possibility that users might sort a region of the spreadsheet by Cut&Paste operations in a straightforward, but somewhat cumbersome, manner. It is not possible to implement the infinite set of possible combinations of Cut&Paste potentially forming a suboptimal Sort Region action. Rather, we can define a common pattern that matches most situations in which the user sorts a region using Cut&Paste (see interchange operation in Fig. 4.25).

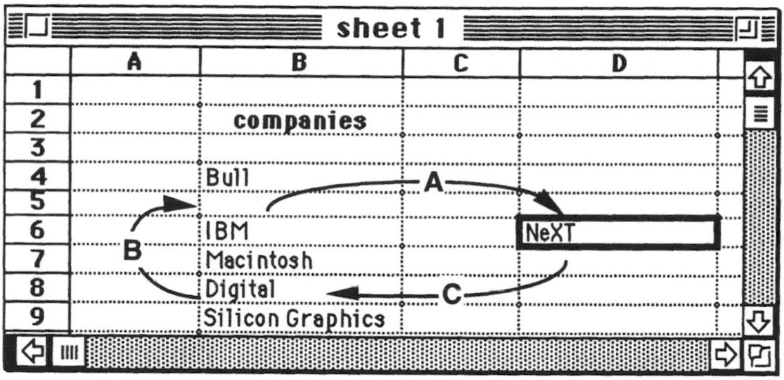

Figure 4.25: Interchange operation in Excel.

If A, B, and C are three Cut&Paste actions, each moving entries from a source to a destination cell, it is possible to describe a common interchange operation requiring:

$$\text{source}(\,A\,) \quad = \text{destination}(\,B\,)$$
$$\text{source}(\,B\,) \quad = \text{destination}(\,C\,)$$
$$\text{source}(\,C\,) \quad = \text{destination}(\,A\,)\,.$$

Requirements like these can be transformed into run-time constraints on the source–destination relationship of the Cut&Paste actions. PLANET has to verify that the current parameter values of all Cut&Paste actions satisfy the constraints. The parameter values of the Cut&Paste action are defined by the parameters of the underlying Choose Region subactions.

The freegoing example illustrates just one very simple model of a suboptimal Sort Region action. To get a generally useful version, we have to modify the given interchange operation model. Most often, users do not exchange single cells, but regions of cells. The source–destination dependencies must be enlarged to regions. To formulate constraints on regions, an intersection operator can be used. Furthermore the relationship between destination(A) and source(C) should be weakened. On the other hand, it is profitable to restrict the operation B to transferring a region of cells within a column or within a row. In our experiments, a pattern like this catches most Cut&Paste-based Sort actions and rejects the nonSort actions.

6.4.4. Interfaces to the Knowledge-base

PLANET provides facilities for describing action plans in the ways just discussed. It uses a special syntax to give the structural description of an action plan:

        structural-action-plan-description ::=
    ((<alternative-1> <alternative-2> ... <alternative-n>)
    <action-plan-identifier>
    <used-by-list>)

Each list is divided into three parts:

(a) The first part enumerates all alternatives (each alternative is an enumeration of the subactions that make up the alternative). Which subactions serve as cue actions is indicated by being followed by 1 or 0.

(b) The second part is the action-plan identifier of the defined plan.

(c) The third part lists all plans in which the defined action plan occurs as a subaction.

The description of the module Declare Database is:

    (((Choose-Region 0 Set-Database 1) (Name-Region 1))
    Declare-Database
    (Extract-Lines-Of-Table Search-Lines-Of-Table   Delete-Lines-Of-
    Table))

The set of all action plan definitions has to be constructed like a Backus-Naur-Form(BNF)-grammar. Every action plan (each expression) can be reduced to a sequence of elementary operations (atomic expressions).

The time- and content-oriented constraints can be declared and defined in the form of Boolean LISP functions. To formulate typical constraints, PLANET provides a library of predefined basic functions. There is a general syntax for defining a constraint as a Lambda function stored in a slot of the plan:

(defparameter <action-plan-identifier>
    ´(lambda (<lambda-list-parameters >)
        <lambda-list-body >))

Function definitions like this overwrite functional slots in the node objects representing an action plan. If the node is activated while spreading impulses up, this functions will be called to check the constraints they involve.

6.5. Inference System

The inference process implemented in PLANET is based on a structure of data objects, each of which is able to receive and send impulses to other objects. A net generator transforms the structural plan description into nodes and links between the nodes. Each plan description corresponds to one node of the inference net. The links between nodes represent the action–subaction relationships. The nodes are implemented as data objects and the edges as pointers connecting the objects. The data structure is used as an impulse-conducting system to spread initial impulses bottom–up. The initial impulses are triggered by elementary events. As subactions combine to form more complex actions, the impulses are gathered and trigger new impulses, which are sent up-level.

6.5.1. Basic Data Modules Used to Implement the Action Graph

The PLANET system is based on a direct implementation of the binary And/Or graph formalism. The nodes and edges generated to represent one action plan define a data module. Each data module contains a central object, several collectors, and one distributor (see Fig. 4.26). The central node represents the goal of the action plan, the collectors define the alternative ways to reach this goal, and the distributor defines how the plan can serve as a subaction in other plans. The objects inside a data module are connected by pointers. Data modules are the basic entities of the net. They are connected by pointers between collectors and distributors.

In the first pass, the net generator transforms each given description into one module. A second pass connects the collectors and distributors of the modules. The time- and semantic-oriented descriptions given as a set of Boolean functions define reserved function slots in the collector objects.

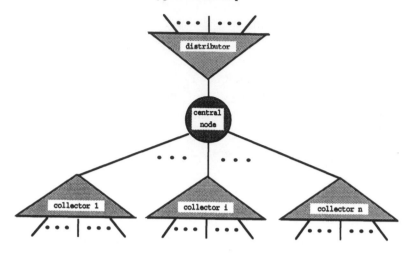

Figure 4.26:   General structure of a Data module.

The graph, constructed as described, is stored in the static knowledge-base and is used to guide the inference process. Both the pointers connecting the data modules and the pointers connecting the objects inside the data modules act as conductors for the impulses, which are spread by a recursive procedure.

6.5.2. Using Spreading Activation as an Inference Mechanism

Each event activates the net at its input nodes. These initial activation impulses spread bottom–up. The collectors gather impulses sent by lower level distributors. The spreading activation process is implemented by a recursive procedure starting at the input nodes, branching at the distributors, and returning at the output nodes. The term *activation impulse* is a metaphor to express the fact that the flow of control is turned over from one data module of the graph to the next. If the collector has been activated over all incoming edges, the impulses are combined in a new impulse. The new impulse will be sent via the central object and via the distributor to all higher level connected collectors. Leaving one data module, the new impulse (recursive function call) holds the information regarding:

- Origin: Pointer to the completely activated collector (executed alternative).
- Sequence value: the maximum sequence value of impulses documented at the activated collector. Sequence values of elementary action executions are defined by the point in time at which they are executed.

The procedure may combine impulses at a collector and generate a new impulse only if the current sequence numbers and parameters meet the collector-specific constraints, which are stored in a special function slot of the collector.

Structure and Use of Activation Instances

The collectors store the activation impulses that have already arrived. Impulses are stored in sets of activation instances. An *activation instance* (see Fig. 4.27) is a field of data that contains a storage cell for each incoming arrow of the collector (number of cells equals number of subactions).

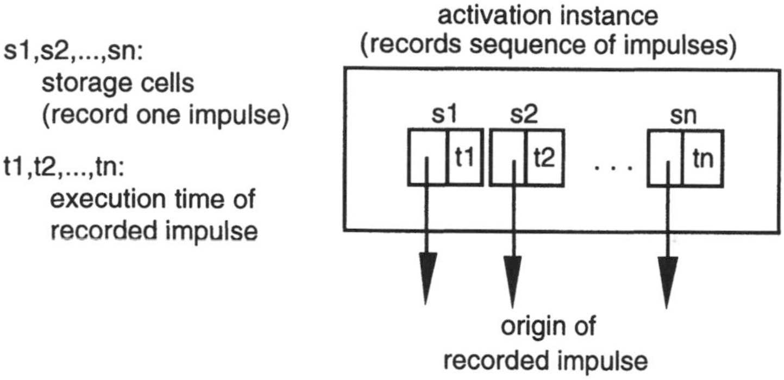

Figure 4.27:   Structure of activation instances.

A cell can only be given a value by an impulse arriving on the corresponding edge. The time and origin information carried by the impulse is stored inside the cell. Any partly defined instance can be completed by impulses arriving later. When an impulse arrives, the constraint functions defined for the collector have to be executed. Only if the impulse does not conflict with an impulse already recorded in the activation instance may the impulse define the instance. A completely defined instance determines a new activation impulse. The origin of the new impulse is the instance just completed, and its sequence number is the maximum of sequence number of all impulses combined in the instance. The integration of the new impulse then proceeds up-level in the same way.

Storing the origin pointers generates a branching chain of activation instances. Each valued cell points to the activation instance where this subaction execution is stored in detail. Again, instances will point to the more specific subaction executions (see also Fig. 4.28). The set of all activation instances defines the content of the dynamic database.

By looking further down along the branches, the inference component can reference all intermediate plan executions and all events on which the plan execution is based. This may be necessary in order to check semantic constraints (see Section 6.4.3). Further on, the branching chain enables the process of evaluating and analyzing the user's actions to access all (sub)executions of a plan when help is requested. Because the recognizer is a component of a help system, it has to be able to represent and manage plan executions that are incomplete at the beginning (prefix), at the end (suffix), or at any other position in the plan. This is guaranteed by storing all legal combinations of impulses at a collector.

Because each plan execution is represented by an instance, the inference process can easily be reset to a previous state. If an Undo event occurs, this can be done by just deleting the instances that have been generated in consequence of the event. We limit the recognizer to managing a variable but limited number of Undos. References to instances built up by the last commands are stored and can, therefore, be subsequently eliminated (i.e., undone).

Generation and Storage of Activation Instances

At a collector, all legal combinations of activation impulses have to be stored. Therefore, most collectors will have several activation instances. The procedure Store organizes the instances belonging to a collector into a storage tree, called a *history tree*. Using a tree structure requires more storage than storing linearly, but it simplifies and speeds up the access and management of the data considerably. Each storage tree has a root node that is always an empty activation instance, the *empty root node*. To store an impulse in the tree, the procedure Store will be called with the Inst parameter assigned to the empty root node.

```
procedure STORE ( IP : impulse ; INST : instance )
if ( IP wants to define an already defined cell in INST or
        IP fails to meet a constraint of collector of INST
) then
                return
    else
            do with  NEXT_INST = first successor of INST   to
                        NEXT_INST = last  successor of
INST
                        STORE ( IP ; NEXT_INST )
            end do
            NEW_INST := DUPLICATE ( INST )
            NEW_INST := NEW_SUCCESSOR ( INST )
            INSERT ( IP ; NEW_INST )
            return
        end if
end
```

The procedure checks to see whether the storage cell corresponding to the impulse in the instance is free and if the time and semantic constraints would hold in a potential combination of the new and already documented impulses. If one condition is not satisfied, the impulse can be documented neither in this instance nor in one of the successors. Otherwise, the procedure calls itself to document the impulse in all successors of the node. Returning, it documents the impulse in this instance node. Thus, it duplicates the instance and inserts the copy into the tree as a new successor of the original instance. The impulse will be inserted into the copy but not into the original. This guarantees that later impulses that are also able to define this storage cell can operate on the same instance. In this way, all legal combinations of impulses will ultimately be generated.

Figure 4.28 illustrates three windows showing parts of history trees of Plans A, B, and C. Plan A contains B and C as subactions. The history tree of Plan A is shown complete. There has been one execution of subaction A at point in time t_1 and two executions of subaction B at points in time t_2 and t_3. The tree of A is the result of procedure STORE. The tree holds all legal combinations of subaction executions that are (Bt_1,Ct_2) and (Bt_1,Ct_3). The references between instances of different trees result from storing the orign pointers (done by Insert).

The storage complexity of the trees is limited in depth, but not in width. If n is the number of incoming edges in a collector, then the maximum depth of the tree is $n+1$. To handle the width of the trees, a branching limit is defined. If the number of branches at a node exceeds the limit, the branch that has existed for the longest time will be cut off.

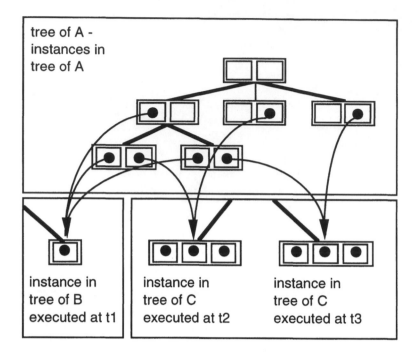

tree of A -
instances in
tree of A

instance in
tree of B
executed at t1

instance in
tree of C
executed at t2

instance in
tree of C
executed at t3

Figure 4.28:   A history tree.

6.6. Evaluating the Inference Results for HyPLAN

Storage trees can be used to evaluate the complete and incomplete executions of plans in the focus of interest. For each execution, the instances show the involved subaction executions and indicate which subaction executions are missing. Each executed subaction can be examined in more detail. The chaining of the instances can be used to reduce each execution to the set of basic elementary actions forming it. Starting from the highest level of action execution in the storage tree, one can determine, at each level, which alternative of the action has been executed. In this way, completely executed action plans can be assessed and suboptimal executions can be determined. Missing subaction executions can directly indicate an Excel usage problem.

Figure 4.29 shows one branch of the structural database with actions and action–subaction relations relevant for Plan A. Dark nodes indicate subactions with complete executions documented.

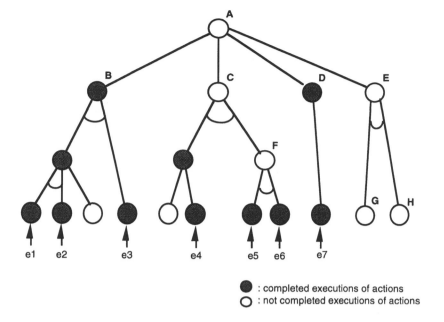

Figure 4.29:   **Partially evaluated plan.**

The execution of action A has not been recognized because the subactions C and E have not been executed. Analyzing the subtree of node E shows that there has not been any execution of the basic subactions G and H. In subaction C, only the subaction F is missing. There has been some execution of basic subactions e5 and e6, but these have not been combined to form F (possibly due to unmet time or content constraints).

The example shows that it is necessary, first, to select an incomplete plan for further evaluation (Plan A). Therefore, we need a component that generates hypotheses about the user's intentions based on incomplete executions. The hypothesis generator we use is simple but effective. A plan becomes an hypothesis if a subaction occurs that is marked as a cue subaction that strongly indicates the desire to execute the plan. Which of the subactions are "cue events" of a plan can be declared in the static knowledge-base. If the plan is hypothesized and the plan is itself a cue subaction of another plan, this second (higher level) plan will become an hypothesis, as well, because of the transitivity of this attribute. As seen in the previous example (Fig . 4.29), hypotheses are important as starting points for the inference process.

Beneath the hypothesis, the completely instantiated actions have to be considered to determine the context-sensitive configuration of help topics. Action plans executed recently are probably of high interest for help offers because the user may have a question concerning the just-executed plan. To have a time criterion, the evaluator manages a dynamic time window of predefined width. A complete execution or hypotheseis can generate an entry in the list of present help topics only if the time of execution lies inside the window (it is not necessary that all included subactions lie inside the time window). With the focus on the selected plans, the user's problem can be related to every sub-action at any level of the given plan's structure. For this reason, the evaluator extracts subactions at intermediate levels.

From the starting point of complete action executions, we have two factors enlarging the selection set. These are the process of generating hypotheses and the necessity of giving help at intermediate levels of complexity. To reduce the selection set, we can take into account time of occurrence and semantic rationales that focus on single subactions, as shown in the example in Fig. 4.29. The action plans finally selected will be mapped to the help topics in HyTASK.

6.7. Open Questions and Prospects

For knowledge-base construction, PLANET provides an interface for describing action modules in terms of structure, sequence, and semantic conditions. Defining action plans by structure requires that the designer have a comprehensive knowledge of the action domain. In order to construct the knowledge base, he has to define the action modules in such a way that further extensions are possible without causing changes in the basic knowledge structure. This task could be supported effectively by a visual representation of the modular structure. It would be very helpful to have a graphical interface that would enable the user to build and modify the knowledge-base interactively.

Determining timing constraints seems to be an easy task for most designers using the syntax provided. In general, the most difficult task for the designer is to take into consideration the semantic plan interactions. Because PLANET manages interactions resulting from overlaps and interrupts, the designer need not worry about that issue. It is, however, often difficult to determine an adequate pattern for recognizing the action under consideration. The designer has to identify constant conditions valid for all action sequences that constitute an execution of the modeled plan but are not valid for executions of other plans. There will rarely be a general syntax to define semantic constraints that can be used in different working domains. PLANET can only provide a library of basic functions for building semantic constraints. Further work would have to be invested in order to develop a satisfactory set of library functions.

Another extension we will develop is an improvement of PLANET's hypothesis generator. A refinement of the concept of cue actions will be implemented by introducing weighted arrows connecting the plan modules. The new version will use real numbers between 0 and 1, instead of the marks 0 or 1 used so far. We hope to solve the problem of missing subactions by having activation functions that collect the weighted activities spread by the subactions. The given weights will be interpreted as certainty factors in the activation function combining arriving impulses. Thus, it will no longer be necessary to differentiate between hypotheses and executions, because every execution will, to a certain degree, be handled as an hypothesis and evaluated by its computed activation value. The activation values of the instances will be compared in order to decide which plans are most likely intended by the user. There are mathematical models of activation functions (e.g., Krishnapuram 1992; Yager 1988) that can be used to express hypothetical human reasoning methods. Fuzzy logic (see Zadeh 1988a, 1988b) provides a mathematical model for expressing vague linguistic terms more adequately. We want to apply fuzzy techniques to model uncertain knowledge about objects and plans. A fuzzy extended version of PLANET will form the basis of a medical support system to confirm diagnoses of congenital heart diseases.

7 . Generalized Criteria for Adaptive Hypermedia Support

Based on our experiences in researching, designing, and implementing HyPLAN, we believe that we can make some generalizations about hypermedia help systems:

a.   Support the user in finding the relevant help.

One of the most basic requirements for adaptive hypermedia help systems, in general, is that they have to support the user in assessing the value of the offered help items in relation to his help demand. This holds true for various stages of the help search, such as entering the help environment or following connections leading to more in-depth explanations of a topic.

Help demos should enable the user to decide early on the relevance of offered help to his actual goal. By providing short descriptions of achievable goals related to a topic, we have to invite the user to match the help being offered with his own goals. By focusing on specific tasks, the user can more easily identify gaps in his understanding and start or continue to look at relevant demos. Our evaluations showed that three- or four-word descriptions related to task goals had the best orientation effect. This is important for the design of the entry level (i.e., the Browser), as well as for link presentations (i.e., buttons) within help demos.

The Browser, especially, supports judgments of the user according to the level of his misunderstanding. A command selected as potentially crucial by PLANET may not be unknown to the user, but necessary preliminaries of the command or its functional embedding in more complex task units may be obscure to him. The Browser is a good example of how hypermedia presentation techniques can be combined with plan-recognition concepts to provide good orientation in a help environment.

    b.   Provide redundant help access and explanations.

To enable users from diverse work contexts to find appropriate topics, redundancy is a necessary and valuable feature of hypermedia help systems. HyTASK, for example, provides the following ways to find information about the Set Criteria command, once the user has chosen the intuitive command Find of the Data menu, but without setting appropriate criteria:

- It appears in the Browser as a topic related to the Find command, as per the topic graph.
- It is mentioned in the text on the help card explaining the Find command and accessible there via hypertext link.
- As the Find tutorial film starts Set Criteria appears in the list of buttons forming links to prerequisite topics.
- The hot list of the Navigator contains a topic named Set Criteria.

Redundancy of explanations employing different presentation media (text, voice, graphics, and animations) is a useful way to let users become aware of important topics, even if they are pursuing other questions. As our evaluations showed, real users pick up on relevant information in mostly unforeseeable ways.

    c.   Focus the development and maintenance of the help system on notoriously troublesome features.

The selection of topics that can be taken into account during the design of a support environment for a complex task domain is necessarily restricted. The range of possible difficulties that users might have with an application cannot be foreseen, and therefore cannot be covered in advance by any technical system. A reasonable design concept, therefore, has to confine support to notorious and crucial mistakes. Thus, help explanations, such as verbal demonstrations and verbal comments, should mainly refer to typical and therefore expected difficulties.

In order, however, to provide a broad spectrum of valuable (multi-level) help, a lot of (mostly domain-related) knowledge acquisition and design effort is necessary. This is even greater if the information presented is to be made adaptive to different work contexts. Support environments built by external developers, therefore, may turn out to be fragmentary or mostly irrelevant for specific target domains.

To guarantee contextual validity of support and help access, (groups of) users in the real target domains should be responsible for the selection and treatment of crucial problems. A maintenance and extension concept that shifts the design of help modules into the system use stage seems to be a reasonable attempt to cope with the otherwise conflicting claims to be contextually informative, exhaustive, and valid. Taking into consideration the need to recognize the help demand and to effectively communicate pertinent information, it can be assumed that the necessary competence can most often be found or developed within a target group. The design of new help demos is made easier for non-specialists if they can exploit a basic set of example topics applying a copy-and-modify strategy. Besides good examples, this requires transparent tools and guidelines.

d.   Avoid active support in favor of passive context-sensitive help.

The fact that most cues for a particular help demand have a vague character rules out active offers. Passive conceptions of context-sensitive help are more consistent with the restrictions imposed by conceivable implementations. In all tests, even human experts were very seldom able to spot difficulties according to the content and level of (mis)understanding without clarification dialogues about the problem at hand. Taking the human model as the most advanced example of cognitive evaluation, we can assume that active one-shot help proposals instigated by a plan-recognition module are an unrealistic goal (Moore 1989). As the Wizard of Oz studies showed, the most that can be achieved by analyzing a record of user actions is to narrow down the possible interpretations of the user's probable intentions and problems. Every reasonable recommendation of help therefore can be seen, at best, as a cautious selection of topics matching the identified user actions.

On the other hand, to come into play at all, such systems require appropriate user expectations. In our tests, users were likely to address the HyPLAN system by the "Help" key after they had been made familiar with the concept. They became more and more confident when its value had been experienced. Using the support environment, the users developed quite realistic expectations and attitudes about the help system. This is an important feature of long-term usability. Although it cannot really be shown by way of single session mock-up tests (Wizard of Oz), we saw that active systems in this sense are likely to be quite inconsistent between different contexts. Without feedback dialogue as in face-to-face situations, active suggestions were often seen as interruptions and were rejected. This holds especially if the user is following a (possibly wrong) idea about how to solve his task. In that respect, error messages are enough to trigger the willingness to receive instructions. A simple technical concept, which we did not implement in HyPLAN, for providing a smooth form of pertinent help access would be click-sensitive buttons in error message windows.

e.   Make the presentations of linked help adaptive.

In HyTASK, the animated demos and the links between them are static. Thus, click buttons are placed at those points in the sequential presentations of task contexts where violations of pertinent requirements may cause trouble. Often, however, the potential relevance of more information is not generally determinable, but depends on the way a task is performed. Therefore, having presentations that can be adapted according to the user's actual work steps, as they are recognized by PLANET, would be a further improvement in the link idea.

References

Allen, J.F., Miller, B.W. (1989):
The rhetorical knowledge representation system (Tech. Rep. No. 215).
Rochester, NY: Computer Science Department.

Berger, F. (1990):
Ein massiv paralleles Modell eines Planerkenners. Diplomarbeit
Informatik, Universität des Saarlands, Fachbereich 10—Informatik.

Bogaschewsky, R. (1992):
Hypertext-/Hypermedia-Systeme: Ein Überblick. *Informatik-Spektrum ,
15,* 127–143.

Bronfenbrenner, U. (1977):
Toward an experimental ecology of human development. *American
Psychologist, 32,* 513–531.

Chun, H. W. (1986):
A representation for temporal sequence and duration in massively parallel networks: exploiting link interactions. *Proceedings of the AAAI,* pp.
372–376.

Dengler, D., Gutmann, M., Hector , G. (1987):
Der Planerkenner REPLIX (Memo Nr. 16). Universität des Saarlandes,
Institut für Informatik.

van Dijk, T. (1985):
Semantic discourse analysis. In: T. van Dijk (Ed.), *Handbook of
Discourse Analysis: Vol. 2. Dimensions of Discourse.* London:
Academic Press, pp. 103–136.

Duchastel, P. C. (1990):
Examining cognitive processing in hypermedia usage. *Hypermedia, 2,*
221–233.

Ferrara, A. (1985):
Pragmatics. In: T. van Dijk (Ed.), *Handbook of Discourse Analysis:
Vol. 2. Dimensions of Discourse.* London: Academic Press, pp. 137–
158.

Fischer, G., Morch, A., Hair, C., Lemke, A., Bernstein, B., Stevens, C.
(1988):
*Explorations in the design of intelligent support systems and innova-
tive user interfaces* (Report). Boulder, CO, University of Colorado at
Boulder, Department of Computer Science.

Flanders, N. A. (1970):
Analyzing teaching behavior. Reading: Addison-Wesley.

Gottschall, K., Mickler, O., Neubert, J. (1985):
Computerunterstützte Verwaltung. Auswirkungen der Reorganisation von Routinearbeiten. Schriftenreihe "Humanisierung des Arbeitslebens", Band 60. Frankfurt/M.: Campus.

Gould, J. D. (1988) :
How to design usable systems. In: M. Helander (Ed.), *Handbook of human–computer interaction.* Amsterdam: Elsevier, pp. 757–789.

Graumann, C.F. (Ed.). (1978):
Ökologische Perspektiven in der Psychologie. Bern: Huber.

Grunst, G., Oppermann, R., Thomas, C., (1991):
Intelligente Benutzerschnittstellen. *Handbuch der modernen Datenverarbeitung, Heft 160.* Wiesbaden: Forkel-Verlag, pp. 35–47.

Hayes-Roth, B. (1988):
Making intelligent systems adaptive (Report No. STAN CS 88 1226). Stanford, CA.

Hill, W.C., Miller, J.R. (1988):
Justified advice: A seminaturalistic study of advisory strategies. In: E. Soloway, D. Frye & S.B. Sheppard (Eds.), *Proceedings of CHI'88: Human factors in computing systems.* Reading, MA: Addison-Wesley, pp. 185–190.

Jussen, H., Grunst, G., Dorn, J., Prinz, S., Kaul, T. (1985):
Interaktionsmuster im Gehörlosenunterricht, Opladen.

Kempke, C. (1986):
The SINIX Consulant: Requirements, design and implementation of an intelligent help system for an UNIX derivative (Bericht Nr. 11). Fachbereich Informatik, Universität des Saarlandes.

Kobsa, A., Allgayer, J., Reddig, C., Reithinger, N., Schmauks, D., Harbusch, K., Wahlster, W. (1986):
Combining deictic gestures and natural language for referent identification. In: *COLING'86: Proceedings of the 11th International Conference on Computational Linguistics,* pp. 356–361.

Koomen, J. (1989):
The TIMELOGIC temporal reasoning system (Tech. Rep. No. 231). Rochester, NY: Rochester, Computer Science Department.

Krippendorff, K. (1980):
Content analysis: An introduction to its methodology. Beverly Hills: Saga Pblications.

Krishnapuram, R., Lee, J. (1992):
Fuzzy-set-based hierachical networks for information fusion in computer vision. In: *Neural Networks* (Vol. 5), Elmsford, NY: Pergamon Press, pp. 335–350.

Kuhlen, R. (1991):
Hypertext: Ein nicht-lineares Medium zwischen Buch und Wissensbank.
Berlin: Springer.

Miller, B.W. (1990):
The RHET plan recognition system 1.0 (Tech. Rep. No. 298).
Rochester, NY: Rochester, Computer Science Department.

Miyake, N. (1982):
Constructive interaction (Report No. 113). San Diego, CA: CHIP.

Moore, J.D. (1989):
Responding to "Huh?": Answering vaguely articulated follow-up questions. In: *Proceedings of CHI'89,* pp. 91–96.

Nielsen, J. (1990a):
The art of navigating through hypertext. *Communications of the ACM, 33,* 296–310.

Nielsen, J. (1990b):
Hypertext/hypermedia. Boston, MA: Academic Press.

Nilsson, N.J. (1980):
Principles of artificial intelligence. Palo Alto, CA: Tioga Publishing.

Oevermann, U., Allert, T., Konau, E., Krambeck, J. (1979):
Die Methodologie einer "objektiven Hermeneutik" und ihre allgemeine forschungslogische Bedeutung in den Sozialwissenschaften. In: H.G. Soeffner (Ed.), *Interpretative Verfahren in den Sozial- und Textwissenschaften.* Stuttgart: Poeschel, pp. 352–433.

O'Malley, C.E.O., Draper, S.W., Riley, M.S. (1985):
Constructive interaction: a method for studying human–computer–human interaction. *Proceedings of INTERACT'84,* London, pp. 269–274.

Palmiter, S., Elkerton, J. (1991):
An evaluation of animated demonstrations for learning computer-based tasks. In: S.P. Robertson, G. M. Olson, J. S. Olson (Eds.), *Proceedings of the CHI'91: Reaching through technology.* Reading, MA: Addison-Wesley, pp. 257–263.

Phillips, M.D., Bashinski, H. S., Ammerman, H.L., Fligg C.M. (1988):
A task-analytic approach to dialogue design. In: M. Helander (Ed.), *Handbook of human–computer interaction.* Amsterdam: Elsevier, pp. 835–857.

Quast, K. J. (1991):
PLANET - Planerkennung mit aktivierten Handlungsnetzen. Sankt Augustin: Gesellschaft für Mathematik und Datenverarbeitung, GMD-Studie Nr. 195.

Quast, K. J. (1993):
Plan recognition for context-sensitive help. In: *Proceedings of the 1993 International Workshop on Intelligent User Interfaces.* New York: ACM Press, pp. 89–96.

Rehbein, J. (1980):
Hervorlocken, Verbessern, Aneignen: Diskursanalytische Studien des Fremdsprachenunterrichts. Bochum: Mimeo.

Rosch, E. (1978):
Principles of categorization. In: E. Rosch, B.B. Lloyd (Eds.), *Cognition and categorization.* Hillsdale, NJ: Lawrence Erlbaum Associates, pp. 27–48.

Schwab, T. (1988):
Beschreibungstechniken für zusammengesetzte Operationen zur automatischen Planerkennung. Universität Stuttgart, Institut für Informatik, WISDOM-Forschungsbericht FB-INF-88-02.

Schwab, T. (1989):
Methoden zur Dialog- und Benutzermodellierung in adaptiven Computersystemen. Dissertation Universität Stuttgart, Institut für Informatik.

Sinclair, J. M., Coulthard, R.M. (1975):
Towards an analysis of discourse. London: University Press.

Spenke, M., Beilken, C. (1989):
A spreadsheet interface for logic programming. In: *Proceedings of the CHI'89,* Reading, MA: Addison-Wesley, pp. 91–96.

Suchman, L.A. (1987):
Plans and situated actions. Cambridge: Cambridge University Press.

Suchman, L.A., Trigg, R.H. (1991):
Understanding practice: Video as a medium for reflection and design. In: J. Greenbaum, M. Kyng (Eds.), *Design at work: Cooperative design of computer systems.* Hillsdale, NJ: Lawrence Erlbaum Associates, pp. 65–90.

Waterworth, J.A., Chignell, M.H. (1989):
A manifesto for hypermedia usability research. *Hypermedia, 1,* 205–233.

Winkler, D., Kamins, S. (1990):
Hypertalk 2.0: The Book. New York: Bantam Books.

Whiteside, J., Bennett, J., Holtzblatt, K. (1988):
Usability engineering: Our experience and evolution. In: M. Helander (Ed.), *Handbook of human–computer interaction.* Amsterdam: Elsevier, pp. 791–817.

Wright, C. E. (1990):
  Controlling sequential motor activity. In: D.N. Osherson, St. M.
  Kosslyn, J. M. Hollerbach (Eds.), *Visual cognition and action: Vol. 2.
  An invitation to cognitive science.* Cambridge, MA,: MIT Press, pp.
  285–316.

Yager, R.P. (1988):
  On ordered weighted averaging aggregation operations in multicriteria
  decision making. *IEEE Transactions on Systems, Man and
  Cybernetics, 18,* pp. 183–190.

Zadeh, L.A. (1988a):
  Fuzzy logic (Report No. CSLI-88-116). Berkley, CA: University of
  California.

Zadeh, L.A. (1988b):
  Dispositional logic and commonsense reasoning (Report No. CSLI-88-
  117). Berkley, CA: University of California.

Chapter 5
Configurative Technology:
Adaptation to Social Systems Dynamism

Michael Paetau

1. Introduction

The discussion of technological realizations of system adaptation exemplifies a category of discussions in the field of computer science concerning two basic, opposing paradigms. First, there is the *active paradigm* expressing the attempt to make the systems as "intelligent" and "autonomous" as possible, to delegate more and more of users´ tasks to the machine, and to relieve users of much of the cognitive effort of having to match the system´s features with their work requirements. Second, there is the *passive paradigm*, which focuses on using human problem-solving potential and assigns to technology only the role of a flexible tool to support human action. This paradigm includes the quite conscious expectation that users make a great effort to intellectually appropriate system features (i.e., to learn to use the system to the fullest). The psychological and sociological investigations carried out within our project were intended as a contribution to the identification of the consequences for users and their work of realizing each of these concepts and to the evaluation of the consequences according to human criteria.

These objectives confront us with all the methodological problems that have been discussed for several years in the context of the debate of technology assessment (see Paetau 1990). Sociological studies, in particular, have suffered because questions about the potential consequences of using a particular technology cannot be answered without consideration of the socio-organizational context, and it is precisely this context that so often cannot be anticipated, or at least not adequately, in the early phases of technological development. However, purely on grounds of ecological validity we did not want to restrict ourselves to laboratory tests with prototypes of adaptable or adaptive systems. In order to realize a "prospective design requirement" we opted for the analysis of *Leitbilder* and concepts. With this approach, we wanted to take the existing knowledge of a pioneering technology that is currently being developed and relate it to an evolving socio-organizational context (still only perceptible as tendencies) and then draw conclusions about future technical development projects. We were particularly interested in developing a concept of socio-technological adaptation that does not—unlike the debate on adaptivity and adaptability—start with various technical approaches, but proceeds from real cognitive, social and, organizational requirements.

2. Reality's Own Dynamism as a Design Problem

In the course of discussions on designing a socially acceptable technology greater importance has recently been attached to criteria, such as flexibility, adaptability, and configurativity. The related idea of a "soft" technology that can be adapted and moulded to changing conditions "on the ground" (i.e., within the field of application) has gradually developed in computer science and has led to several new concepts of varying breadth. As an issue of software ergonomics it was first discussed in the 1980s under the concept of "individualizability" and found its way into various guidelines and evaluative frameworks for the human-oriented design of application software (see ISO 9241, Part 10; DIN 66.234 (8); VDI 5005; EVADIS; and others). Here, discussion first centered on the question of adapting application systems to personal characteristics of the user, such as his level of expertise in using the system, individual preferences for utilizing the system as a result of individual differences in style of acting, thinking, and learning; and so on. In this discussion, the research domain expanded step by step to problems of tasks and organizational contexts. Consequently, in software ergonomics we now speak of four mutually influencing factors (and, thereby, four design factors): the human, the technology, the tasks, and the organization.

The first phase of the flexibility discussion in software ergonomics was typified by a widespread design orientation on the tool character of technical systems. Despite the fact that there were quite a few authors who emphatically pointed out that computer systems are essentially different from classical tools (see Wingert & Riehm 1985), it was common to see the development of human criteria for software design from the perspective of a user with a single tool appropriate to his specific work. As an alternative to the societally dominant "management's perspective," the intention was, thus, to develop the objectives of workplace humanization and to oppose the usual business strategies of systematic rationalization (see Nake 1986). From this standpoint, a system can be regarded as flexible if it "is structured in such a way that the user can employ the same system to perform his work efficiently even when his tasks have changed, that the user can perform a particular task in various ways which he can choose on the basis of his changing level of expertise in each case and his current proficiency, and that different users with different backgrounds and experience can fulfill their tasks in alternative ways" (VDI-Richtlinie 5005 1990, p. 18).

This perspective, which is geared to the individual workplace and entails an understanding of the concept of flexibility as "individualizability", became problematic as workplace computers were increasingly linked up. Now, with networks evolving everywhere, the traditional analysis criteria were called into question, although they still held for the individual workplace. By the time computer science moved on to developing complex systems for cooperative work (CSCW or groupware), however, it was clear that the tool perspective had been overtaken. Now, it became necessary to focus on the whole socio-organizational context as the frame of reference for human work and on its relevance for technology design. In the field of software ergonomics this insight won in two steps. First, by taking work tasks into account the classic human–machine dualism was extended and relieved "cognitivism" as the predominant theoretical approach. The user could no longer be reduced to an *information-processing being*. But after the user had also been recognized as a *task-processing being*, it became increasingly difficult to ignore his nature as a *social being*. In the field of software ergonomics increasing consideration was now given to the socio-organizational context (see Oberquelle 1991), and the research domain extended more and more. Because the problem posed by technological design was increasingly recognized as an issue that encompasses the totality of the social and technological conditions for the labor process, and was thus perceived as a process of comprehensive socio-technological innovation, it was no longer possible for the criterion of flexibility to be reduced to individual technical machines or components. The basic question became how technical working environments could be designed to manifest such high flexibility that they would not, themselves, determine and specify work structures, but would permit the technology to follow the behavior of social systems that are always adapting to changing environmental conditions.

What is remarkable in the discussion about the "question of design" (Winograd & Flores 1986) is not the fact that technology must be consciously shaped, for that is nothing new—especially, of course, in the engineering sciences and organizational theory. Rather, the decisive point is that there was a change in our understanding of design. The design interests of computer scientists had previously concentrated largely on technical artifacts; at best, they understood their design assignment as only touching on the question of user interfaces, while generally regarding themselves as a neutral body of experts in disputes over uses and effects. Computer scientists are now reflecting more and more on the fact, that in the pursuit of technical solutions, they provide far more than what is supposedly just the means for achieving objectives laid down by others. Technology design is—whether consciously articulated or not—always work design, too. Thus, technically adequate design can only be achieved by addressing the goals for which technology is to be produced and by looking at the environmental context in which technology is to develop its practical effectiveness. In computer science, the question of technology can, therefore, no longer remain restricted to the question of an appropriate means–ends relationship, but must include the broader issue of the objectives themselves.

There is another central point in the current discourse: should, or can control of complex sociotechnical processes take effect from within or from without? In the field of computer science, this problem has been a subject of fierce debate, at least since Winogradand Flores' publication in 1986. In social and organizational sciences, discourses with nearly the same subjects occur. For a long time, the social sciences attached insufficient value to the multilayered and multicausal nature of the processes of organizational change[1]. Work design could only be grasped as a planned activity directed at predetermined goals and, usually, as an intervention into a social system, mostly led through external change agents. In contrast, questions are now increasingly being asked about the internal social relations in a company that influence the course of technological-organizational innovation. This necessitates the analysis of systemic characteristics that were previously largely excluded; in fact, their empirical exclusion in connection with the fulfillment of organizational tasks or design processes was a functional aspect of this discipline.

The central point of the debate is turning on a new understanding of systems. Whereas the old cybernetic concept viewed control as external, the new concept of system is more concerned with internal control. Central to this approach is the concept of self-organization, developed by several scientists, in various (natural and social) scientific disciplines and from the standpoint of very different scientific traditions. [2]

This discussion has met with enormous interest in the sphere of practical application. It is not the organizational structure but technology that is currently regarded as the main obstacle to realizing new concepts of production and service, because technology is proving too unwieldy for the necessary processes of self-organization. Thus, in contrast to the earlier situation, in which the application of the latest technologies was seen as bringing competitive advantage, the position has now been reversed: Companies hope to become more competitive by turning to new organizational concepts.[3] For these to work,

---

[1]   An exception can be seen in the "organization development" approaches that have been used for many years now. These concentrated at a very early stage on the inclusion of those affected and the development of their competence of action (cf. the concept of "organizational development",   French & Bell 1977).

[2]   Cf. Maturana & Varela 1990 (autopoiesis); H. v. Foerster (second order cybernetics); H. Haken (synergetic); and E.v. Glasersfeld (radical constructivism). In all of these approaches questions of self-reproduction, self-reference and self-organization are focused on as central problems.

[3]   Cf. the current debate on "lean production" (Womack, Jones, & Roos 1990).

however, a technology is required that is not yet available. The characteristics that are currently lacking and are most sought after are found, mainly, in configurative systems capable of being flexibly incorporated into a permanent process of innovation that is regulated by the organizations themselves.

3. Self-Organization of Social Systems and Distributed Intelligence

Attempts to design technical or social systems assumes, explicitly or implicitly, that systems can be influenced at certain points in a desired manner. If one succeeds in finding suitable criteria and methods and obtaining sufficient knowledge about the magnitude of disturbance factors (e.g., undesirable side effects), then changes can be brought about in the way one wishes. For many years, the predominant view regarding such regulatory processes has been based on ideas developed in systems theory that primarily aimed to exert a targeted influence on particular structural components of systems. The new debate in systems theory—above all in response to the works of H. von Foerster, E. von Glaserfeld, Maturanaand Varela, H. Haken and others—has broken with this idea, inasmuch as the structure of systems is now conceived of as a moment that is constantly moving and adapting to changing environmental conditions. Systems are no longer tied to particular component structures or configurations. They are made up of the coherence of evolving and communicating processes; in these processes temporary structures are created that are not characterized, for instance, by a particular state of equilibrium, but precisely by their constant movement. This movement allows order to be maintained, enabling the system to support and reproduce itself.

The new perspective on the genesis and development of structures does not, however, apply to all systems, but to those with the specific system characteristics of living systems. Living systems have the capacity for self-referentiality, self-production, and self-organization (Roth 1987). Luhmann has—heedless of Maturana's reservations (in Riegas & Vetter 1990)—transferred this perspective to social systems (Luhmann 1987), a development that has fallen on fertile soil in the organizational sciences (see Türk 1989).

Movement, life-world (Lebenswelt), and social action are concepts with which the recent literature of organizational theory has been operating and, in trying to describe sociotechnological changes, they seem to me to be of great importance. In Maturana and Varela's theory, a system's organization arises from the specific way in which the individual components communicate. They produce emergent features of the total system. "Organization refers, then, to the relations between the elements of a unit that functions (interacts) as a whole with the characteristic properties of the respective class of units"

(Riegas & Vetter 1990, p. 334). The organization of living systems is described as "autopoietic" by Maturana and Varela, because the interaction of all cell elements enables the self-reproduction of the cell as an autonomous unit and self-reproduction ensures its continued existence. In self-reproducing, the structure of the system can change, but the organization remains invariant (Riegas & Vetter 1990).

The current upheaval in organizational theory—as in the field of computer science—expresses the difficulty in describing adequately the complex multi-causal relationsbetween different factors using traditional scientific methods. Until well into the 1970s, the dominant notion was that a particular structure evolves in organizations on the basis of particular system–environment relations, that is, a structure that is adequate to this specific relationship. It was thought that with an optimal relationship between context variables, and structure variables one could determine an ideal organizational structure that would decisively improve its efficiency. New approaches in the field of organizational research have broken with this assumption, just as they have broken with the notion that rationally motivated means–ends relationships are decisive for the workings of an organization. Today, more attention is paid to the social processes of group and norm forming and to the development of social networks than was the case in earlier work. Türk (1989), for instance, wrote of a "resociologization" of research into organizations.

I do not wish to try to provide a detailed presentation of the current theoretical debates on organization. However, a number of new arguments are of great relevance to the issue at stake here (Türk 1989):

- Organizations are primarily social systems. Even if one agrees that they have an emergent structure, that rules are laid down within them and, that these rules are also actually followed (more or less), organizations are first and foremost constituted by life-worlds. In other words, the action of organization members is, to a high degree, shaped by the existence of specific organizational structure and subcultures. Such subcultures can have both integrative and dissociative effects. What is significant is that social action and the associated communication are the medium for the self-reproduction of the organizational entity.

- This insight means that the analysis of the generally accepted objective structure—and I do not wish to question its existence either—takes second place to the analysis of emergent interaction processes and habits. The specific objectives, values, and models that predominate in organizations are not, for instance, given focal points toward which organization members direct their actions; rather, such focal points are generated through social actions within the organizational structures. This means that, contrary to earlier assumptions, we may regard them not so much as variables that explain the action of organization members, but as phenomena that are, themselves, in need of explanation. The same is true of behavioral rules, which are generally overtaken by everyday practice before they are revised as formal rules.

• In saying this, it is also clear that the stability of organizations cannot be achieved by attaining a particular state of equilibrium; on the contrary, organizations actually maintain themselves through a process of permanent adaptation and movement, including constant structural change. Demands change, tasks differentiate, patterns of cooperation shift, authorities are established or called into question, and groups appear and disappear. These processes produce structural changes that define the identity of the organization. Protection of the organization through structural variations, then, is what characterizes organizations as "living systems" undergoing an evolutionary process.

This perspective also leads to a new challenge in defining human–machine relationship. After all, we are dealing here with a relation between two classes of system that differ fundamentally in their characteristic features. Unlike human beings, machines are not capable of reproducing themselves, and, moreover, they are subject to external, human-determined, purposes. According to Maturana, these systems are allopoietic[4]: a feature that is fundamental to the "identity" of technical and social systems and to the related possibilities for getting the two systems to relate to each other. On the basis of this paradigm, it is, strictly speaking, no longer possible to refer to a human–machine system or to a sociotechnical system. Clear-cut system boundaries must be drawn between social and technical systems. Their mutual relations are established via a common medium through which the two systems are linked, in the mode designated by Maturana as structural coupling.

4. Human–Machine Relations as Structural Coupling

In this new understanding of the human–machine relationship the qualitative difference between the two systems does not lie in certain structural features but, rather, in the different characteristics of the self-reproductive organization of each system (Schmid 1987, p. 27). Thus, what constitutes the identity of a system is the organization, which constantly reproduces itself by changing its form. This happens in constant interaction with the system environment. Man and machine, society and technology, represent each other's "system environments". They are structurally coupled (Luhmann 1991, p. 108).

The concept of structural coupling casts a different light on the assumption common in the traditional sociology of technology that technical systems

---

4   I therefore depart from Maturana's concept in the distinction he draws between two types of machines: allopoietic and autopoietic. For Maturana, *autopoietic* machines are identical to living systems.

have particular effects on social systems. Because technology and society represent two "closed" systems existing alongside each other, there cannot, in the conceptual framework of the new systems theory, be any direct cause-and-effect relationship in the input-output sense. Strictly speaking, one can no longer speak of technological impact. Technological and social systems exist simultaneously alongside each other so closely that they appear to merge.[5] Causal relations can only exist within the same system. The interrelationships between various types of systems is, in principle, so complex that attempts to regulate all potential causes to ensure a particular effect must be regarded as an illusory enterprise. In the area of human–machine communication, for example, it is hardly possible to ascribe definite causal factors to specific actionparameters. Personal features, such as skill, behavioral and learning styles, and so on, play a role alongside the characteristics of the work structure (functional differentiation), the organizational conditions (systemic and subsystemic differentiation) and technical features (variety of the available application systems, user interfaces, etc.).

For the technological system the social system is an environment: Technology is implemented by the social system and is required to respond to the latter's demands (organizational processes, pattern of work tasks, etc.) with appropriate functions. Being allopoietic, technlogy is dependent on explicit interventions on the part of the system environment (i.e., by the social system). These occur in the form of certain system development measures, adaptive actions, and configurations undertaken by social actors. From the point of view of the social system, on the other hand, technology is part of the system environment. "Society adapts to the pure presence of technology" (Luhmann 1991, p. 108). Thus, technology, in general, and information and communications technology, in particular, have become frameworks for everyday human action; they are part of culture. Structural coupling exists in the inability to move backward to another (lower) stage in the development of the productive forces. When technological systems break down, they lead to disturbances in the social system, but as long as they are connected through loose-coupling,

---

5   This problem is, for example, posed by attempts to achieve a holistic design of work and technology. The statement that "technology design is work design" is justified if as one seeks to draw attention to the fact that any use of technology leads to certain irritations in the sphere of organizational and social systems, but it does not determine specific effects. The extent to which changes take place in organizational design depends essentially on how the social system itself treats these irritations on the basis of its own structural determination. Despite the deep interpenetration of work and technology, the social and technical systems remain completely separate from each other, because both system obey their own laws, at any point in time.

society is able to cope with such disruptions. In the case of minor disturbances, a few changes occur in the normal societal processes whereas major disturbances can, under certain conditions, require a process of structural adaptation.[6]

These adaptive processes are provoked by changes in the system environment. A comparison of technological and social systems shows a qualitative difference in the way adaptation actually takes place. Whereas social systems—understood in Luhmann's sense as autopoietic systems—reproduce all their elements themselves and, thus, in the course of their history, take a path of constantly self-produced structural change, technical systems cannot do this, because they are allopoietic systems. In other words, machines are extrinsically controlled and regulated systems, requiring regular interventions from outside, but—and here lies the difficulty—there are no fixed optimal parameters to guide such external regulatory action. Unlike clear-cut means–ends relationships, the precise result of regulation is not knowable from the start, but only emerges later through the unfolding of the dynamics of the social system. For software engineering this means that concepts must be found that enable technology to adapt to the living dynamics of the social system.

5. Systems Design Caught Between Context Dependency and Application Neutrality

Although the adaptation of technology to human beings is a demand that was first raised long ago by engineers in relation to their development approaches, it is a demand that is still repeatedly expressed and continues to provoke fresh discussion in the light of changing technological possibilities. As long as the machine is understood as an allopoietic system, adaptation can only be conceived of as something that takes place through the intervention of engineers in the course of a cyclical development process. Because we are concerned with adaptation to the real dynamics of the world of work, the prerequisite is close coupling between technological development and work design. The close connection between application and technological development has long been a focus of interest in the field of computer science. The principle has been quite frequently advanced, for example, that the starting point for developing software systems should be work design. However, there are several obstacles

---

6   These adaptation processes are triggered by changes in the system environment. It is contended that man's freedom to create a world according to set objectives lies in the possibility of influencing the system environment itself and maintaining the correlation of both interlinked systems by shaping the (technical) system environment into a specified form.

to realizing this idea: after all, even if technological development and work design are viewed as inseparably linked, they are still not one and the same. They are (generally) executed by various actors, at various times, and in various places, and the greater the social, spatial, and temporal distance between the two, the more difficult it becomes to find an adequate description and specification of the set of real-world problems to be addressed by the software. This is especially apparent in the case of software developed for an anonymous market (standard software) or software developed for specific basic research problems, where it is impossible to forecast the precise field of future application; in this case, software engineers are too far removed from the application stage of their evolving software. In such cases, the normal process of feedback (as found in, say, software development rooted in a particular field of application) is not available to software engineers or only occurs after great delay through the mediation of the market, so they have no way of approaching the future application system other than to engage in anticipatory modelbuilding. Such models are—and this has been no secret in computer science for some years now—in principle inadequate. Indeed, they are inadequate in several ways: with respect to the individual task to be described, with respect to the activities in which that task is embedded, with respect to the persons who carry it out, and with respect to the structures of the organizations in which the activities are being done.

Application development therefore finds itself caught between the construction of systems that correspond as closely as possible to desired applications and provide good support for concrete tasks, on the one hand, and the development of standard systems that have a broader scope and support a large number of possible types of tasks, but in so doing have to generalize from the concrete features of a task on the other. Systems that closely fit their tasks are preferred in practice because they do not require the user to spend a great deal of time and effort adjusting and adapting the system. Their disadvantage is the limited ability of such systems to be used for different tasks or for changing tasks. This usually means that the user must—depending on how complex his activity is—work with various systems. He is often required to learn a number of different system functionalities and user interfaces. Where this is the case, the standardized systems have a great advantage: They put together a complex system that can, in principle, be used by the user for all of his tasks. However, experience shows that most of these systems turn out to be suboptimal and still require appropriate adaptation, necessitating the presence of expert consultants or of user expertise in handling instruments for adaptation, which are generally very complex and, in our experience, can hardly be mastered by the average user (Karger & Oppermann 1991).

From this situation arose the idea of having adaptation carried out autonomously by the machine itself. This approach views the machine as a sort of autopoietic system which, thanks to a number of technical features, undertakes the adaptation by communicating with the user's work environment. It was the starting point for the SAGA project. Before examining these two technological approaches, however, I must first explain a few basic structural characteristics of system adaptation.

6. Four Perspectives on System Adaptation

Reasons for constructing flexible (adaptable or adaptive) systems may be advanced from four different perspectives: from the technical perspective, from the viewpoint of the user, from standpoint of the tasks to be processed by the user, and from the perspective of the organizational context in which these tasks are carried out.

6.1. Technology

From the software engineering perspective, system adaptation mainly serves to bridge the distance between system development and system application. It is well known that one of the most basic difficulties in the development of application software is anticipating the potential situation for using and applying the embryonic product accurately enough to meet the requirements of task-appropriate and user-adequate support for the target group. These requirements are more likely to be fulfilled the more closely software engineering is linked to the field of application, the more narrowly work activities are circumscribed, and the more precisely the special work styles of the targeted users are known. However, a trend has been observed in recent years -at least in the clerical sphere—that tends to point in the opposite direction: As a result of the widespread use of PCs, a growing proportion of software is standard software, and there has been a decline in systems developed within the application context (e.g., by companies' own data-processing departments or by specialist software firms contracted to provide customized company solutions). The reasons are primarily economic: Although far-reaching specification of software in terms of a particular type of task and user can improve a product's usefulness, it also increases the production costs. Any software produced for an anonymous market (i.e., not in a definite application context) necessarily faces the basic dilemma of having to be universal enough to be used by as many persons as possible in a particular class of application, while, being able to be employed in response to very special user requirements and needs that are not even known to the software engineer and that, as a consequence, either he cannot model or he can only model inadequately. System adaptivity is intended to meet this underlying problem of all software engineering.

6.2. People

The second strand of the argument in favor of the adaptability of technical systems has emerged in the context of ergonomic studies. In particular, the findings of industrial psychology concerning the personality-advancing effect of greater degrees of freedom in the performance of work (Hacker 1978) and the basic doubts surrounding the development of work systems geared to the idea of "one-best-way" (Ulich 1978) have given rise to the demand that information technology systems should not only prescribe a specific mode (conceptually anticipated by the systems engineer) of using and processing information, but should also enable users to develop individual problem–solving strategies to support them in discovering or creating new possibilities for action, in influencing and transforming traditional forms of task–solving and in adapting dynamically to new situations (e.g. environment conditions). In this way, the usefulness of information technology systems cannot be measured merely by the degree of efficiency with which given tasks are fulfilled, but by the potential for transforming intervention in the structures of the task-solving process.

The basic question of user-oriented system adaptation is which criteria should be used to specify the adaptation features. Discussions have centered on adaptation facilities geared to the level of experience of a user in handling a particular system (level of expertise), to the style in which the individual processes information (verbalizer vs. visualizer), to the degree of flexibility in a person's actions (flexible vs. rigid), to the degree of self-reflection (reflective vs. impulsive), to the extent of field-dependence or field-independence in cognitive recall, and to the individual's problem–solving style (holistic vs. sequential forms of problem–solving).

One of the most important criteria for classifying and modeling users is their level of expertise. This measurement is derived from various approaches in cognitive psychology. In particular, the works of Johnson-Laird (analogy models), Anderson (production models, Norman & Rumelhart (network concepts), and Minsky (schema concepts) have influenced the discussion. Various user models can be designed on the basis of certain assumptions about the differences in the cognitive structuring of knowledge and different forms of access to this knowledge (Eberleh 1989).

In 1990 Karger reported on the literature on the user models contained in various implementations of adaptive systems and distinguished the following dimensions of modeling:

(a) Orientation on firmly defined user types or on individual users (stereotypes vs. individual models).

(b) Specification of models through explicit stipulation (system, designer, user) or through system-side derivation on the basis of user behavior (explicit vs. implicit models).

(c) modeling on the basis of long-term or short-term characteristics, such as the user's general field of interest, expertise or task field, or situation-specific user behavior when solving a quite specific task (long-term versus short-term models).

To categorize users by criteria such as those just listed, the systems need variables to determine particular features. AID[7] for instance, utilizes the variables error frequency, help-referral frequency, and frequency of use of the system itself, and specific commands. With the Help of these variables, the system categorizes the user in a user class for which a specific type of support is then provided by the system and for which the appropriate depth and detail of help information or feedback from performed actions[8] is regulated. The goal is to respond to individual differences in user expertise in the handling of a system. In ascertaining individual differences, other systems draw on particular user stereotypes made up of various features, such as learning strategies (e.g. MONITOR or AKTIVIST).[9] These systems compare the actions actually performed by the user with previously defined action plans that are characterized as "efficient", with the object of identifying suboptimal operations and automatically supplying relevant help options.[10]

The assumptions that go into this kind of model construction do not all apply well from the standpoint of cognitive psychology. Karger (1990) pointed out that knowledge cannot be treated as context independent and stable. Whether a command is easy or difficult for a user to remember depends on so many factors that it is difficult to create a uniform user class on this basis. Moreover, there are considerable problems in transferring the learning situation—usually, communication between a (human) teacher and a (human) learner—to a machine-mediated tutorial situation. With these objections in mind, one might conclude that, ultimately the users themselves must be the

---

[7] This refers to the adaptive user interface for an e-mail system based on UNIX. The development was carried out by STC Technology, Essex.

[8] Repeated complaints about poor feedback have been voiced by users of UNIX systems, in particular.

[9] Both prototypes are active and application-independent help systems, MONITOR was developed by the National Physical Laboratory (NPL), Teddington, UK (Benyon, Innocent, & Murray 1987, Benyon & Murray 1988); AKTIVIST at the University of Stuttgart, Germany, by the INFORM project group (Bauer, Herzberg, & Schwab 1987).

[10] A detailed description is given in Karger (1990).

ones who—on the basis of their often-changing personal or work-organization situation—actually decide on the necessary adaptive measures.

6.3. Tasks

Alongside the concept of user adaptation, discussion has also centered on the concept of "adapting to work tasks", which is sometimes explicitly presented as an opposing position.[11] As in the case of user-oriented adaptation, the problem raised here is the choice of criteria by which such adaptation should be undertaken. The attempt to develop models of office-sector tasks (similar to the aforementioned user models) with which a system might control its adaptive behavior or be controlled by the user does, however, seem to me be a very doubtful undertaking.

Efforts to produce a generalized description of office activities usually draw on classifications found in industrial sociology and business administration; these were compiled either to generalize empirical findings on particular trends in the evolution of work and organizational structures, or to offer a rough orientational framework for companies to use in practice. One of the central criticisms of this approach is that in a real work activity the dimensions of content-, form-, and role-specification are intermingled so their meaningfulness as indicators is limited. Gottschall, Mickler and Neubert, for instance, distinguish between a routine worker and a knowledge worker. In a task-based classification Szyperski Grochla Höring and Schmitz distinguished management, specialists, professionals, and assistants. Picot & Reichwald also presented a task-based typology, which, incidentally, is characteristic of almost all of the business administration approaches.[12] They differentiated three task types—ad hoc tasks, specialized tasks, and routine tasks—that are determined by task context. Classification features are the level of problem complexity, the level of complexity of information retrieval, the amount and scope of cooperation with other people, and the general scope of action. Nippa identidied not tasks, but positions. He classified the following types: leadership positions, strategic expert positions, administrative expert positions, and task-processing positions.

---

[11] This was one of the issues discussed at the workshop on "Possibilities and Use of Individual System Adaptations" held in the GMD, Schloss Birlinghoven on January 30, 1990.

[12] An exception is the approach taken by Nippa (1989) in his attempt to transfer the three task types of Picot & Reichwald into a typology of job positions.

In my view, a meaningful description of office activities should distinguish among the following three dimensions:

(a) An expert-domain–oriented dimension, referring to the substantive purpose of work and the specialist skills and qualifications that are needed. Historically, specialized areas have evolved in the different occupational branches. The activities of clerical workers are segmented along these lines (e.g., the preparation of loans in banks, the administration of policies or the processing of damage claims in insurance institutions, etc.).

(b) A functional role–oriented dimension, that is, work activities geared to the smooth functioning of task-fulfillment procedures (e.g., coordination operations), but largely independent of the specialist-content objective of organizational assignments. Here, we can identify a broad spectrum of activities ranging from directing to pure implementation activities.

(c) A formal information-processing–oriented dimension, that is, the pattern of information-processing activities through which the other dimensions are connected (i.e., activities that cope with expert-domain or role-oriented work). In performing these tasks, the people are processing knowledge (their own or external). Information is gathered, generated, stored, evaluated, transformed, modified, and so on.).

Computer systems that try to support all possible combinations of these three dimensions are almost bound to fail. Although the user facilities of such systems might cover all conceivable general cases, they are usually suboptimal for the individual activities. In many sectors the demand for systems suitable for specific tasks has led to the development of highly specialized applications tied to particular expert domains. These systems can certainly be very broad-based in the range of object classes to which they are applied (e.g., processing of texts, graphics, files, etc.), but they only allow one to manipulate objects within the narrow boundaries determined by particular expert domains (e.g., acquisition support for insurance representatives, a warehouse management system, etc.). In these systems the use complexity is reduced to those functions that are necessary to deal with the specialized expert domain. The result is an increase in task adequacy but a restriction in flexibility. Other products are designed with more emphasis on dimension (c), formal information-processing,[13] and leave both other dimensions—the expert-domain and the role orientation—largely unspecified. They concentrate on a few object types (e.g., text, graphics, data, etc.) that can, however, be processed for a variety of specialist-content task types. Because most tasks require different object types—for example, in preparing a conference one needs structured data, continuous text, graphics and so forth—the user is forced to employ several application systems (or highly complex multifunction packages) to do his job. The task adequacy of such systems is limited by the object types that are available. On

---

[13] E.g., database systems, spread sheets, text systems, and so on.

the one hand, this approach is flexible because the system's use is not restricted to predefined specialist-content task-types, but, on the other hand, it increases the cognitive requirements with respect to finding and using system facilities.

The point of an adaptation concept oriented to tasks or task-types is to reduce the aforementioned pressures by enabling a user or user group to make a task-adequate configuration of the system functions.

6.4. Organizations

Modelsthat go beyond the formal context of individual tasks structured by the division of labor and attempt to replicate cooperative work (i.e., the social action associated with task solving) have so far hardly proved useful in view of the real complexity of social systems (Türk 1989). Although the structural coherence of individual tasks may be described in the form of flow-charts, network diagrams, and so on, when it comes to the process of problem–solving in the framework of a social network, such descriptions are not very useful. Yet, it is precisely the process character of organizations that is at the center of the latest research into organizations, especially in all those approaches that take up the new systems theory paradigm, described earlier.

The character of organizations has long been idealized. Max Weber considered organizations as ideal–type embodiments of rational action; seen as goal-oriented planned systems, they were ascribed a (more or less permanent) objectified structure (Weber 1972). In the older conception of organizational theory, deviations from the ideal type were understood as "informal phenomena", which then became the object of supplementary "hygiene" measures designed to guide the action of organization members (e.g., the human relations approach). New approaches in organization theory (from sociological, psychological and business administration perspectives) have since corrected this picture. The assumption today is generally that organizational structures and processes are contingent.[14] Attention focuses more on the following points:

- Organizations are conceived as action contexts with cultures and subcultures. People work in organizations, but they do not only work. They also live and love in them, and dream and fight, and rules are laid down and broken in them. There is order and chaos, and yet (or perhaps for that reason?), they function.

---

[14] I use the term *contingent* here in the general sociological sense (as in Luhmann or Parsons). This should not be confused with the contingency-theory approach, known from business organization theory (e.g., in Kieser & Kubicek 1978).

- The mechanistic view of an organization expressed in organigrams and flow-charts (i.e., the image of hierarchical relations and formal structures) derives from an external standpoint and remains superficial. It tells us little about the real actions and operations through which an organization constantly reproduces itself and evolves.

- The structures relevant to the functioning of an organization can best be considered as being created by the interaction of subjectively interpreted ideas held by working individuals concerning the content of work, the distribution of competence, the ways and means of reciprocal interaction and so on.

- Not only are organizational structures partly the result of goal-oriented, rational planning and action, they are, for the most part, the outcome of internal, situational action constellations (Weltz 1986) in which different interests reciprocally act on one another.

- Rational system behavior cannot be explained by a relation between environmental dynamics and organizational structure. In the recent literature, hardly any organizational theorists still maintain a determinism between particular environmental situations and system behavior. Earlier attempts to identify ideal–type relations between context variables (e.g., industrial branch, market situation, company size, forms of technology, etc.) and internal organizational structure variable (e.g., form of specialization, decision-making centralization, etc.) have largely been dropped.

In view of these characteristics of organizations, considerable doubt attaches to the attempt to devise an anticipatory model description of organizational structures and processes on which technological design can be guided. First, what one can describe are the formal elements of an organization, such as the specialist-content definition of subsystemic units, the formal definition of areas of competence, the hierarchical structures, the tasks laid down in job plans, and the flow of material and files needed for the work process, and so forth. However, this does not depict the actual social process of cooperative work. Central to the special quality of organizational action are, above all, the forms of social integration that are expressed in group relationships, the formation of constellations of influence and power, the course taken by inter-group and intragroup conflicts, the development of action-guiding group norms, the modes of interaction, group solidarity and forms of consensus-seeking and conflict.

The major difficulty for software engineering here is the awkwardness of such—often called "informal"—elements of organizations,[15] in contrast to formal descriptions and specifications. Yet, it seems to me that there is growing recognition that despite the difficulties involved, we should not lose sight of the fact that these elements exist and, indeed, have a constitutive function

---

[15] Recent works tend to speak of "organizational culture".

for the entire organizational process, for it is precisely these elements that increasingly offer a key to optimizing rapid and flexible reaction to changed situations (external market positions, sets of internal company actions, etc.). The transgressing of rules, the breaking out of fixed role expectations by organization members, and the situational taking of responsibility for a particular event (e.g., to prevent an accident) are processes that make it possible for organizations to react to unforeseen situations. Anything that limits this informal sphere and ties down the members' scope for individual discretion and independent action ultimately suppresses the potential for the flexible unfolding of organizations.[16]

Computer science faces a dilemma, because, in contrast to organizations (which, as social systems are capable of self-referential evolution), technology—and the same applies to software—is only genuinely flexible during the development process. By the time it is implemented in the socio-organizational system, technology has already become rigid and can only follow the given rules. Its application primarily strengthens the sphere of formal action in organizations by expanding the scope for action and behavior in that sphere. At the same time, however, it tends to confine informal, spontaneous, rule-breaking, and role-transforming actions (Falck 1991). In other words, the very area in organizations that facilitates autopoiesis is being impeded, namely, the flexible potential for organizational action. The discussion of adaptive systems starts with this problem. What adaptive systems should look like, which system levels and system elements should be encompassed by adaptive features, how complex adaptive tools and methods can be while remaining controllable, and which actors should perform the adaptive measures are all questions for which a number of different ideas have been put forward.

7. Dimensions of Adaptation

Adaptive facilities can operate on various levels of the technical system and have very different points of orientation. As a whole, we can identity four different dimensions by asking the following questions:

(a) To what should the adaptation be oriented? There are two or three possible foci: the user (action style, learning styles, habits, etc.), the task (different types of tasks), and the position of the task embedded in its organizational context (e.g., changing organizational arrangements).

(b) What range should the adaptive measures have? On the basis of VDI guideline 5005 (software ergonomics/office communication), there are four

---

16 These functions are behind the effectiveness of the "work to rule" tactic in conflicts of interest between employer and employee.

different levels: task level, function level, operation level and input/output level (Paetau 1990, pp. 233 ff.).

(c) Who should undertake the adaptive measures? Conceivable modifications are those performed automatically by the system (auto-adaptivity),[17] by the end-user or group of end-users, by the user's own advisors (in large companies usually located in the user service of the central data-processing department), or by the consulting experts of the software producer or systems engineer (especially for major alterations and servicing, which cannot be carried out by the user's data processing department, as is the case with expert systems).

(d) What means can be used for adaptation? In addition to the software engineering tools in use, which, of course, can in principle also be used for adaptive measures (even though their use requires the avalibility of software engineers), one can distinguish between application-internal adaptation tools (e.g., command configurations, batching operations into macros etc.) and application-external means (e.g., generic interface tools).

In terms of their scope, previous adaptation options have concentrated mainly on the low-level human–machine relationship, the operations level, and the input–output level. This is particularly the case with auto-adaptive systems, which have therefore proved to provide only partial solutions. The initial expectations in the modeling and technical representation of user types or task types as a means of producing adaptive options at the higher levels (task level or functional level) have proved too optimistic. The only aspect that can be given an adequate description is the formal and constant structure of definite tasks, and that is not sufficient for automatic adaptation to user work activities. However, the task and function levels are of great significance for those involved in social, human-oriented work design who want to go beyond molecular and merely routine actions. In other words, there is a need for dynamic adaptations whose concrete form can hardly be anticipated, and this requires adaptation tools with easy access, without restricting the adaptation ranges.

8. Results of Interviews

As already mentioned, the project we carried out from 1989 to 1991 looked at various technical concepts that had emerged from discussions within computer science in recent years. It was our aim to analyze them in terms of their technical feasibility and their social benefits. At first, the debate within computer

---

[17] On the technological preconditions for the auto-adaptive version of a system adaptation, see. Krogsaeter and Thomas, and Fox, Grunst, and Quast in this volume.

science was marked by a technical dichotomy: on the one hand, the concept of self-adaptation (also called adaptivity or auto-adaptivity) and on the other, the concept of adaptability. The decisive criterion distinguishing the two approaches is the question, who takes the initiative for making specific adaptations: the human or the machine? This debate took on significance that went beyond the narrow software ergonomics issues, because it expressed an underlying dispute over the future direction of developments in the field of computer science. It essentially addresses the far-reaching question of the degree of "autonomy" that can responsibly be given to technical systems. Two basic and opposing positions can be identified. One focuses more on the attempt to make the system as intelligent as possible, to reduce the user's cognitive effort, and to largely relieve him of his own adaptation work. The other focuses more on using human problem–solving potential, on designing a flexible *tool* also capable of innovate action, and on the opening up of system facilities to human intervention, so that the user is deliberately expected to invest considerable intellectual energy in learning how to use the adaptation tools.

We carried out various studies on the practical significance of adaptive systems for the organization of work. On the one hand, we wanted to find out whether the adaptation options available in some of the systems generally available on the market were actually being used and the reasons why they might not be used. In addition we wanted to determine how far the idea of adaptive systems has been taken up in the development of concepts for work organization.

The study concluded that, at present, adaptive systems play hardly any significant role in work design concepts, even though all the respondents regarded the basic idea underlying such systems positively. On the whole, the use of adaptive tools was seen as too complicated and as encumbered with a number of cooperation problems. With these results in mind, one must ask how far the concept of self-adaptation can go in providing a solution.

Self–adaptive systems should try to support the user in handling a system in the learning and orientation phase. It should draw the user's attention to unfamiliar tools and utilities that he can use for the rapid accomplishment of certain operations he may have previously performed inefficiently or in a roundabout fashion. By offering context-dependent objects and functions, the system becomes less complex, reduced to the size needed by the user for his current actions. If the user makes errorsor becomes confused and thus needs help, the adaptive system should offer him task-related and user-oriented explanations. In the case of user errors that can be clearly interpreted as such, automatic error correction should remove the need for explicit correction.

To achieve this goal, a machine knowledge base operates as a mediator between user and computer. For the machine knowledge base to take on this mediating function, it must have stored knowledge about the user's work context, his tasks, his individual working styles, his mental model of the computer, and his resources, as well as methods of dealing with typical problems found in the user's task field and heuristic assumptions for deriving new solutions. By evaluating recorded user behavior, it should be possible to make the system capable of adapting to the user's tasks and modes of behavior. The user's current actions are examined to see if there are regularities corresponding to previous action sequences. Where correspondence is identified, the system draws conclusions about relevant follow-up actions or action objectives. In this way, computer systems should, as with human communication partners, be able to adapt to expected user behavior.[18]

A number of reservations have been expressed from the user's point of view about the concept of self-adaptation. One criticism is that the user is under "observation" by the system; his behavior is recorded, and, thus individual work profiles can be obtained. Problems of data protection—or, rather, protection of personal privacy—arise because the user is subject to supervision by evaluating levels of authority. This external control option is, say the critics, compounded by a loss of individual control on the part of the user. As result, the know-how acquired through working with the system cannot be applied by the user in a way that allows the formation of a reliable mental model of the system. The knowledge and experience gained would again and again be made virtually worthless by the automatic adaptive performance of the system. The user's competence and control in the work process would therefore be weakened, rather than strengthened. The user would lose his overview of the system's structure and its performance capabilities. Thus, the system would fail to meet the important software ergonomic criterion, that it should conform to the user's expectations (see ISO 9241: "conformity with user expectations").

The technical implementation of auto-adaptive systems has turned out to be more difficult than expected, and all the relevant questions are far from answered. Quite apart from this, however, our current findings show quite

---

[18]   Speaking metaphorically, on the basis of the given set of rules of the programming of machine plan-recognition mechanisms and the logging of current user actions, a computer should certainly be able to "see things from the user's action perspective". However, it can only do that—and this is the key distinction from interhuman communication—on the basis of a particular quantity of similar actions that show a certain regularity. A computer is not able to make an adequate assessment of individual cases by, say, spontaneously identifying the situational context in which a particular interaction takes place. Thus, the ability of a modern computer-based system to recognize action plans is still fundamentally different from what is referred to as "empathy" in interhuman communication. On the basis of this underlying problem, the technical conception was drastically changed in the course of the project.

clearly that the question of how to develop an appropriate adaptation concept cannot only have a technical answer; indeed, the solution is not even primarily technical. In terms of its basic character, system adaptation in complex work contexts must be seen as a social process and not simply as the autonomous actions of individual end users (see section 8.2.). Certain requirements are derived from this understanding aids push the dichotomy of auto-adaptive versus adaptable into the background. To the fore come considerations as to how the existing problems of making adaptive facilities useful can be overcome, in terms of both the cognitive aspects (e.g., greater transparency in access to adaptation tools, better opportunities for explorative learning, etc.) and the co-operative aspects (communicability of adaptation results, suitability of adaptive tools for group processes, etc.). From the perspective of a work-oriented, ergonomic, and socially desirable system design, it is apparent that, at the higher levels of abstraction of human–machine communication (the task level and the functional level), the initiative for adaptive measures lies with the human. At the lower levels (the operations level and input–output level) there are numerous examples in which automatic adaptation appears very sensible, even if there are still unresolved problems here too (see Krogsæter, Oppermann & Thomasin this volume).

To determine the role played by adaptive software in recent design concepts for clerical work, to assess their potential for implementation, to identify possible problems, and to facilitate feedback to the conceptional and technical development work, we held interviews with representatives of various companies and associations. For these empirical surveys, we only used qualitative methods (interview analysis). So as to include the user in our investigation, we talked with experts from the trade unions responsible for the relevant sectors and with a technology consulting unit run by the DGB (German Federation of Trade Unions), which is concerned with both industry and insurance. In total, 20 experts were interviewed: 8 interviews were at insurance companies, 6 at industrial cooperations, 3 at advisory organizations, and 3 at trade unions. The set of topics dealt with in the interviews covered:

- Current trends in the design of organization and work in the respective sectors, especially concerning concepts and practical forms of "holistic office work" from the specialist content point of view (e.g., integration of previously separate procurement sections in industrial administration); from the functional, role-specific point of view (e.g., reintegration of activities previously divided between experts and assistant personnel); and from the formal, informtion-processing–oriented point of view (e.g., the integration of different information processing activities).
- The importance of individual data processing and the efficacy of the corresponding user support provided by the company.
- The development of skills and qualifications concerning the division of function between human and machine.

• The potential uses of adaptive and cooperative systems in the future: for instance, to which persons or groups of persons they might apply, or how they might affect the companies organization concepts.

We originally surmised that the trend toward holistic office work observed in recent years was partly associated with greater freedom for system individualization. This view proved deceptive. Although almost all interviewees recognized the criteria of flexibility and individualizability as a universal ergonomic necessity for software, they also expressed a number of organizational reservations. In generalizing about these reservations, it is necessary to distinguish between the two sectors investigated, insurance and industrial administration.

In the insurance business we generally found a rather positive attitude toward individualized software. However, one must take into account the strong division between processing activities and the work of experts in the central, executive staff sphere (e.g., internal auditing, actuarial mathematics, management organization, law, taxation, etc.). Processing duties were regarded by all the respondents as very standardized and hardly open to individual system adaptation options.[19] In contrast, the experts have far more scope for action. Their spheres of operation require individual data processing. Thanks to their training and educational background (e.g., mathematics), they often have quite extensive knowledge of information technology. In fact, it is not unusual for these higher personnel to make individual adaptation measures themselves or even to write small programs, although this, too, has its limits, and in some cases the trend is now toward restandardization.

In administrative departments of industry, the division between experts and specialized clerical staff is not so clear-cut. The latter's space for individual action is significantly greater than in the insurance sector. Individual data processing is not only done by knowledge workers, but, to some extent, by clerical workers, too. The greater scope for individual action is not, however, expressed in ideas for achieving more opportunities for the individualization of systems. On the contrary, the existing potential to apply individual systems and to modify them to fit one's own requirements is regarded as a dan-

---

19 In the interviews, the experts sometimes referred to "conveyor-belt work" or "factory work" as industrial metaphors. This was primarily the case with the administration of existing insurance contracts, ranging from the drawing up of policies and the amendment of contracts (e.g., risk changes, sum changes, recalculation of premiums, etc.) to cancellations. Here, there is hardly any freedom of action for the clerical staff. The individual work procedures are largely formalized and thus automatized, so that work activity is to a large extent bound by technical software formalism.

gerous tendency (referred to as *wild proliferation*), which the organizational side must stop.

All in all, we can sum up the reservations expressed in the two sectors in the following points:

- *Support Problems:* In the wake of the widespread movement towards individualizationalmost every staff member has his or her own individual system, so it is now hardly possible to provide full support through the central data-processing department. The same applies to assistance offered by individual colleagues, especially in regionally decentralized organizational structures: Even the standard software is no longer uniform.

  "We are only in favor of individualization if support from the advisory service can be guaranteed. In many instances, excessive individualization can no longer be supported by the advisory service and it pushes up the costs. That is currently often the case. Many different systems are available, especially in the IDP area. That is a problem because we can no longer sensibly service everything. So there is now a trend towards higher standardization." (Minutes 5, p. 9)

- *Problems of Staff Substitution:* Finding a substitute for a staff member in the event of sickness or other form of absenteeism becomes more difficult the more closely the working environment is tailored to the employee being replaced. This problem does not so much concern the systems' user interfaces as the individually designed structure of file management and the individually regulated forms of access to particular files, and so on.

  "If everyone starts using his own little system then enormous problems will obviously arise. As for user-interfaces, you can probably solve that problem out via individual profiles and the like. But there are also difficulties with the work itself. You only have to think of filing procedures. Even now we suffer from being unable to get a large group to keep to a specific filing structure." (Minutes 17, p. 12)

- *Problems of Power and Transparency:* Depending on the individuals position in the company hierarchy, worries were expressed about the increasing lack of intelligibility and transparency of work performance in the company, or about the loss of overview of the central data-processing department.

  "In the past we have really learned a lot about clerical processing through molding office work to fit machine processes. In cooperation with the specialized departments, we got around to defining many functions that were previously not at all clear. We have made a lot of processes more uniform. By contrast, in the past every clerical worker had somehow managed to get something done. The customer often received widely different responses and relatively inconsistent decisions on identical issues. The advantage of the whole change to mechanization was not only that we have saved on labor—that, too, of course—but that we have created clear-cut rules for specialized processing work and thus transparency. And I now realize that the problem of such computer assistance is that the machine's ability to react very individually leads to a loss of transparency again. And that is something I don't want to see. We are actually more interested in making procedures more transparent, in having clear rules and uniform responses to the same state of affairs. To this extent I have a rather critical view of such a machine in the clerical processing sphere." (Minutes 2, p. 16)

- *Networking Problems:* Due to networking with other workplaces problems are seen in the further processing of documents which contain individualized elements arising from their creation.

> "We are not pushing that ... very much because it results in a series of islands. When someone then leaves their job and another must take over, he won't know where to start. ... I can see two mutually contradictory concepts there. One begins with the individual: what he needs for his well-being. The other concept starts with the imperative to use the means of production as economically and efficiently as possible. Between these two concepts, you somehow have to find a compromise. The limits will probably be found where an activity moves away from purely individual work and where other staff are affected. This is where rules have to be laid down. At this point, not only is the tool stipulated but also the way of working." (Minutes 18, p. 10f.)

The dominant problem was clearly seen as the provision of back-up for users. Even if there was an extensive transfer of user support to the technical system itself, few of the respondents believed that such support would suffice.

> "A human being needs warmth, and a screen is not particularly warm. I am not certain that help functions will bring about the big breakthrough. Everyone is happy to hear a voice that also offers comfort: 'Come on, it's no problem, we'll manage OK.' etc., rather than the user always having to read the cold prompts: 'More questions?: Then press next-page.' Of course, it is also a question of the quality of the help system, but I cannot imagine personal support being completely replaced." (Minutes 12, p. 11)

As a general concept, the possibility of having information technology with greater adaptability or even configurability from the department level right through to the workplace was seen positively (and this was also the view of the trade union representatives questioned), but on the basis of previous experience, the experts were rather skeptical about the prospects of effective implementation.

> "I think the user service department has a very important function. And it can be divided up and located at many different points. If users start to do everything by themselves, there will be dreadful chaos. From the data privacy aspect, too, there must be people who keep an eye on what is actually going on and who are at hand with expert advice, but they have to be on site and not only at the central office. We will probably end up with a combination of the two: On the one hand, the technology will develop so that it becomes easier to handle, especially for occasional users and, on the other, there will be some people on site who can go a bit deeper into technical problems. The idea that the systems will one day be so designed that they will be able to do everything is, in my view, an illusion." (Minutes 20, p. 7)

There was quite a variety of responses to questions of practicality. In some cases, the idea of system adaptation was rejected as impractical. In other cases, the companies tried to find their own ways of using the existing skills to develop a cooperative solution. The organizational structure, especially the relations between central and decentralized organizational units, plays a major role in determining how the various companies reacted to the contradiction described. Characteristic differences were found between the two business sectors analyzed. For example, insurance companies are strongly centralized in the administration of contract-related data, but strongly decentralized in customer service. To overcome the difficulty of making a satisfactory user service available to the decentralized areas,[20] forms of user support were developed that involve cooperative elements. Here, use is made of the potential offered by a certain number of qualified personnel with a specific technical propensity and the relevant skills (so-called "computer freaks"). Such additional skills, which staff often acquire through time spent on home computers as a leisure activity, are deliberately encouraged and developed in the required direction through company training schemes.[21] These staff, then, become the ones colleagues

---

[20] A telephone advice service is available to these staff around the clock.

[21] "I would like to limit individual assistance to basic matters that can be generalized. For individual problems, I would like to make use of the freak potential. After all, the freaks get involved with the systems. Why should this potential remain unchanneled and simply evaporate? It can be utilized by the company." (Minutes 14, p. 9)

should approach in the specialized departments or groups (known as coordinators) and, depending on their individual know-how, provide advice, "first aid", or even interventions to modify a system.[22]

9. Configurativity: Self-Design in a Design Network

The concept of *configurativity* that I now define is linked to three points in the previous discussion:

* First, to the objective necessity of making information technology systems within the application sphere accessible for acts of modification, adaptation, configuration, and individualization.

* Second, to a company's real problems of implementing these requirements in practice.

* Third, to specific problem-oriented proposals developed in some companies to tackle, at least temporarily, the contradiction between objective requirements and the lack of realization; proposals that respond to immediate needs, rather than following explicit concepts.

The approach proposed here arose out of discussion on the flexibility of information technology systems. *Flexibility* is conceivable in various technical forms. It can, for instance, be understood as diversity. In this concept, the use of a system is facilitated in accordance with certain previously anticipated, modeled and technically constructed alternatives. This solution would, however, depend on the anticipatory modeling and specification of cooperative activities (i.e., on an endeavor that has already been called into question). The concept of *configurativity* attempts to overcome this difficulty by seeking to extend the *living phase* of technology (i.e., its development phase) into the application phase. This attempt has been also been undertaken with the aid of other concepts, ranging from end-user programming with fourth-generation languages, and application-internal programming and macro language right through to forms of adaptivity. None of these approaches, however, has yet been able to live up to the expectations placed on them. This is partly due to the fact that the technical solutions are still inadequate, that learning and handling the adaptive tools is too difficult for the end–user in the specialist departments, and that there are some cooperative problems associated with adaptation measures (Friedrich 1990; Paetau 1990).

---

22 "In addition to the central support structures (...) there are also decentralized coordinates. (...) And as a rule someone with a problem first goes to his colleague and, if he can't help, then he comes to us." (Minutes 4, p. 10)

The configurativity approach proposed here takes up on various ideas found in previous approaches while being distinctive in key aspects. It differs on two fronts: it differs, first from the software ergonomic concept of individualization and adaptivity and, second, from the more recent software engineering concepts like those expressed in the prototyping approach (Budde et al. 1992) or the evolutionary software engineering approach (Lehmann 1982; Smieja & Mühlenbein 1991). The configurativity approach is most closely related to the concept of user programming (Döbele-Berger et al. 1988), though with major qualifications. I do not assume that users who participate in the design of the socio-technical systems in which they are situated must learn and apply a programming language. I believe methods should be made available that allow the actors to manage without such wide-ranging technical skills. In the following, I want to clarify the concept, first by showing how it differs from other approaches and then by presenting its positive definition with a more precise explanation of its adaptation and configuration dimensions (subject, methods, range, and orientation points).

9.1. Distinctions in Relation to Other Concepts

9.1.1. Configuration Versus Software- and Organization Engineering

The concept of a "design hierarchy" (e.g., in Hacker 1987 or Kubicek 1989) does not, of course, only describe various successively abstract levels of the work situation, in which the content of each level builds on the previous, and each level is accorded a larger or smaller range of influence on the working individuals. Rather, it is also based on a sequential design process that passes through the individual levels, step by step. Thus, this concept includes the hierarchical specification of certain aspects of the organization *and* the technology. In general, the following levels are specified: work tasks, division of labor and combination, work procedures, human–machine division of functions, working equipment, hardware, and software (and these, in turn, in the sequence: functionality, user interface). In short, we can identify the following basic assumptions behind this design model:

- Design processes have a sequential and hierarchical structure.
- Adequate methods are available to obtain sufficient information for a correct, complete, and unambiguous description of the individual design steps.
- This information can be fed into model constructions and plans capable of fixing a technical organizational system of goals.
- This goal system can claim validity for a specified time period.

- To adapt the technology adequately to changes inside and outside the organizational structure, one must, after a certain time, initiate a new design process, which is subject to the conditions listed.

9.1.2. Configuration Versus Prototyping

In contrast to the prototyping approach, we do not work on the assumption that a completely specified goal system must be established. Rather, our concept is more like the approach taken by Lehmann, who spoke of "evolutionary software engineering". However, *configuration* differs from both of these approaches with regard to the subject. We envisage a self-design process involving the active participation of actors with different qualifications and skills. Actors with software-engineering expertise can be involved, but need not be. That places certain demands on the available methods, tools, skills, and qualifications in an organization. A universally valid approach cannot be defined, because there are a variety of solutions depending on economic sector, size, and level of organization. In decentralized organizations, individual expertise plays a greater role than in centralized organizations, which generally have their own advisory services.

9.1.3. Configuration Versus End-User Programming

The concept of end-user programming has been under discussion since the mid-1980s, especially in connection with the development of fourth–generation programming languages, and some pilot projects have been carried out (Döbele-Berger et al. 1988). Based on the rather successful developments in the industrial work context (especially with workshop programming of CNC machines), attempts have been made to transfer these ideas to the office sector, but so far, without success. One must ask where the causes of this obvious discrepancy may be located: (a) in the specific content of work (working on physical materials vs. processing information; (b) in the skills expected of the user (technical vs. commercial competence); or (c) in the specific experience-based expertise (experience in the transformation of action goals using a technical tool vs. a more communicative impact).

9.1.4. Configuration Versus Individualization

The concept of individualization maintains that an individual end–user can adapt the system available at his own work station. Discussion of this idea has concentrated chiefly on aspects of the user interface and less on issues of system functionalities. The central issue has been the scope for individualizing the input–output interface and the interactive interface, such as changing keyboard layouts or the display of menus (e.g., "short-menus" or "full-menus"). The tools for this kind of individualization (see the detailed treatment in Simm and Oppermann this volume) may be of two types: application-specific (i.e., as a component of the user interface of a particular application system and, accordingly, only usable for this application system) and, *application-neutral* (i.e., for all, or at least many, of the application programs available on a particular computer). Such application-neutral adaptive tools can be provided at the operating system level (e.g., the control field settings for Apple Macintosh, which one can use to specify how the mouse movements are translated to the screen, the time interval for the mouse's double-click, the volume of acoustic signals, the color display, culturally specific parameters, such as units of measurement, etc.), but also as universally applicable supplementary programs (e.g., "QUICKEYS" for determining key combinations to execute certain commands, or "DIALOGEDITOR" to change the appearance of dialogue boxes). In many applications, there are options for adapting operations through operation linking (creating macros), for object combination for the same operations (print chains), or for the joint management of individual objects.

It is, above all, the range of potential interventions that distinguishes the concept of configurativity from this sort of individualization. Configurativity includes the functions themselves and, from there, impacts on the task level. The beginnings have already been made in some of the latest systems, although they are still bound up far too closely with particular applications. They can be seen, for example, in systems where it is possible to define special objects (e.g., presentations in WINWORD). What is lacking is the possibility of defining task-specific functions and reducing or expanding the complexity of the systems in accordance with the type of task being performed.

9.2. Actors of Configuration

One can discern a shift of emphasis regarding the question, "Who performs the adaptation?" The previously dominant concept of single user adaptation will give way to a cooperative concept. The cooperative character remains fundamental to configuration actions, even when one can foresee that, in the future course of technological development, new shifts will always be taking place between those adaptation measures that can be made autonomously by a user and those that come about as the result of cooperative processes. With every technological improvement to the human–machine interface it will be possible to extend the autonomous interventions of the end-user; but, at the same time, new adaptation facilities are always being developed that can only be executed on a cooperative basis. Thus, one can insist on the cooperative nature of adaptation processes as a basic principle that—as in any other cooperative context—also encompasses individual performance.

The end-user and advisory staff can be involved in this cooperative process at different levels, as can several end-users working autonomously. The form, in which the adaptation measures are performed (with or without the help of advisory personnel and with what range) depends on a series of concrete conditions, such as the substantive range of specialist activities of the respective users, their actual forms of cooperation and concrete cooperative requirements in the workplace, their position in the company, their level of expertise in using the relevant systems , the organizational structure of the company, the available technology, and more besides. All these pertinent conditions vary so much from company to company that they cannot be generalized in a single, broad adaptation concept. As a consequence, all considerations of how to provide technical support for system adaptation must proceed on the basis of a cooperative process in which the system adaptation features are typically incorporated, while its concrete course cannot be anticipated. Thus, technical adaptation concepts must focus on adaptive tools that are capable of supporting a social process of innovation.

In this sense, we must concentrate on ideas of how to overcome the problems as we try to usefully apply adaptation facilities, taking into account both the cognitive aspects (greater transparency of access to adaptive tools, better possibilities for interactive and explorative learning, etc.) and the cooperative aspect (communicability of adaptation results, suitability of adaptive tools for group processes, etc., Paetau 1991).

9.3. Methods

From the software engineering perspective, the approach described here is based on object-oriented thinking. It foresees a free configurability of objects or object classes (the free structuring of text, language, object and pattern graphics, structured data), basic applications (object processing, managing, transportation and communication, organization aids) and functions (processing methods), that users would, have to be able to control without programming skills. Through the configuration of special objects (including the necessary basis applications and functions) that are adequate for a particular set of concrete tasks, it should be possible to perform system adaptation measures on a task-related basis. The users themselves shall decide just how adequate the adaptive options are. They then have the possibility of reviewing the configuration by bringing their own practical experience to bear and, if necessary, modifying it. By combining a chain of objects, they also have the option of linking typical sequences used to solve problems. This means they configure for themselves not only ideal–type objects but also ideal–type (and always reversible) action processes.

The class formation of the various modules, as understood here, should not follow the traditional object classes, such as text, graphics, data, and so on, but rather, elementary task types. In other words, a module brings together task units, which in one sense are bigger than usual (i.e., they cover different object classes), but in another sense are smaller (they only carry part of the tools and methods associated with traditional object classes). The class formation of the modules must be empirically determined. It must not be too specific, because the user would then find it too complicated to combine the modules. Yet it cannot be too universal either, because that would prevent it from offering efficient task support and ultimately lead to the same dilemma faced by today's standardized systems, which have to be stuffed with a large number of adaptive facilities.

The analyses carried out in the SAGA project have clearly shown that system adaptation constitutes a cooperative problem, not only in relation to the execution of adaptation measures, but also with respect to the outcome.[23] These cooperative problems must be taken into account in a realistic concept. The configuration tools must be developed in such a way that (in addition to the software ergonomic requirements for transparency, cognitive intelligibility, etc.) they contribute to initiating a step-by-step learning process in the group, which gradually does away with the need for support from data-processing experts. This is achieved by broadening the existing qualification base (i.e., the various detailed abilities and special skills found among group members) to create an overall set of skills and expertise for the group as a whole. The aim must be to start a social group process by transferring the distribution of the collective expertise onto the shoulders of all the group members and starting the necessary synergetic process that releases this expertise when a concrete problem needs solving. To this end, one must ensure that the old structures in the relationship between data-processing experts and work specialists do not reproduce themselves in the form of a sort of microcosm (e.g., by certain work specialists with special data-processing know-how developing into mini-advisors within the group).

---

[23] See the operational problems discussed earlier concerning support and advice for users, substitution and replacement in the event of illness or other forms of absenteeism, networking with other workstations, and power over and transparency of the work performance in a company.

The adaptation tools must have groupware character and support those areas of cooperative work that can break the rigidity of "collective thought-worlds" that exist in a group (i.e., the internalized division of tasks between data-processing experts and work specialists) and change these thought-worlds. I am thinking here, above all, of forms of system support for explorative and communicable processes in which the results of configurations can be made transparent and can be easily exchanged with other groups or individuals. This also applies to the question, habitually arising in organizations, of which system elements have to be standardized and which can be freely configured.

Because the configuration of basic applications, objects, and functions should not require programming knowledge on the part of the user, standardization has to be set at a relatively high level. That can happen in the form of preconstructed modules that the user only has to put together. The assembling of the functionality is achieved with the help of a tool-box giving the users access to the object classes, basic applications, and functions provided by the system; users can then make configurations and record them in special (task-oriented) system versions.

9.4. The Problem of Competence

The problem of demanding too much of users by expecting them to perform complex configurations is a central concern of all the concepts of system configuration, adaptation, or individualization. I believe that user expertise is not activated by calling up knowledge that has been developed at some point and is then, in principle, available. In the case of highly complex systems, like an integrated office system, new learning processes are needed on a continuous basis especially where the system features in use are not routinized. Learning takes place intentionally, during problem–solving, error discovery, and correction; incidentally, during casual observation of work processes; and, of course, via tutorials and exploration. A high level of system complexity can prevent this learning effect, because understanding and localization of a problem is made more difficult and navigation difficulties arise, when one explores the system. If learning is to be supported as an interactive process, the totality of all the elements to be learned must be limited to the immediate problem-solving procedure being tackled in a particular context. In this sense, the fostering of competence cannot be understood as a kind of repair job on suboptimal work, but rather, as guidance toward the system functionality relevant for the user′s actual problem-solving situation by enabling and supporting an active confrontation with the system that also allows for cooperation with others.

In contrast, the conveying of knowledge via tutorial learning takes place in the form of "canned" units in which the user has to leave the ongoing work process. The transfer of the identified abstract problem solutions into one's own working environment has to be performed by the users themselves. This type of learnig (learning by knowing) requires different cognitive efforts from the type of learning described earlier (learning by doing). Active learning for computer work has been increasingly emphasized in recent years. Empirical studies show that active learning leads to high efficiency in the correction of planning errors and is superior to tutorial instruction (see Karger 1990). On the basis of these findings, there has arisen the concept of a gradual expansion of functionality made dependent on the user's know-how (see Carroll & Carrithers 1984), and this can be seen in rudimentary form in the Macintosh environment, where users choose between short and full menus. However, such concepts have not been realized in task-specific form. I argue that real support is provided for the learning process when the functions appearing on the user interface are directly geared to the problem area connected with the immediate task at hand. Only then will the user (trying to remember what he has already learned or perhaps forgotten) venture along the path of exploration, whereas an interface offering the entire functionality will tend to scare the user off. For this reason, task-specific reduction of complexity is a key element for system adaptation.

10. Conclusions

One of the most important finding of general validity to come out of the SAGA project is that the discussion, of how to adapt technology to humans, which has already been going on for quite some time, cannot be continued along the lines it has so far taken in computer science. As long as software engineering sticks to the notion that the characteristics to which technology is supposed to adapt can be described, specified, and modeled, every advance in technological design will remain at least one step behind man's actual practical requirements. This means that software ergonomics must break free of the familiar ideas inherited from traditional ergonomics. Although it was still possible for traditional machine-oriented ergonomics to, as it were, measure man and his work and derive ideal–type constructions and norms by which individual machine elements could be constructed, this is no longer the case in the design of computer systems. Computers are a means of mechanizing mental activities and socio-organizational structures. These elements are subject to substantially stronger dynamics of change than, say, the parameters of man's physical features, such as the length of an average arm or sitting postures. In the development of human mental work, we are dealing with an evolutionary process whose range and speed cannot be compared to the evolution of the physical characteristics of human work.

From what has been argued here, it may be concluded that, in general, *configurability* within the application field will be a central design criterion for computer-based systems built to match the contingent and complex character of organizations. Organizational arguments derive, above all from new insights into the contingency and internal dynamics of organizational systems. In contrast to earlier (hierarchical) conceptions, it is now generally understood that organizations are social networks with complex and dynamic configurations, so that their behavior is not subject to a simple determinism. By grasping the character of organizations as self-referential systems that can autonomously generate constant structural changes, we reveal a key distinction vis-à-vis technical systems: Once technology has been implemented in the respective sphere of application and the development process has been brought to an end, the prospects of further modification to adapt the system to the dynamics of a living organization are generally very poor. Thus, the interest in adapting technical systems is basically concerned with the attempt to extend the development phase into the application process, to shorten the distance between development and application in a user-controllable form so that technical systems can still be shaped in the application field.

From the findings of our empirical investigations, we can formulate certain demands for the design of configurative systems. First and foremost are considerations as to how the existing problems in the utilization of adaptive features can be overcome, and this concerns both the cognitive aspects (greater transparency of access to adaptive tools, more possibilities for interactive and explorative learning, etc.) and the cooperative aspects (communicability of adaptation results, suitability of adaptive tools for group processes, etc.).

The concept I have briefly presented here focuses on the following points:

- Regarding the problem of the focus of adaptation and modification features, users should be able to gear system changes to the different tasks that occur in the spectrum of problem-domain–oriented and functional-role–oriented activities. In this way, the organization as an environment that affects individual problem-solving action will also be taken into consideration.

- Regarding the actor of adaptation (i.e., who carries out the system adaptation measures), the approach pursued here assumes that system adaptation is only partly performed as an individual process: In terms of their basic character, adaptive measures are part of a process of cooperation that integrates the individual's actions. The cooperative nature of the adaptation process must, therefore, be incorporated in design considerations from the very beginning.

- Regarding the means by which system adaptation measures are undertaken, users should not be expected to perform programming themselves, but should be provided with the appropriate means to configure their work environment in accordance with their respective tasks. This configuration potential should be available at every level of human–machine communication. In other words, in addition to the individualization possibilities that have already been realized in rudimentary form at the operations level and the input-output level, configuration must become possible for all available objects or object classes, for all basic applications and for their related functions.

Unlike the systems available today, this concept is distinguished by the following features:

- Users can do their own configuration and filing of object classes provided by the system (freely shaping text, language, object graphics, pattern graphics, structured data) and the required basic applications (object processing, administration, transport, communication, organization aids) in special (task-oriented) system versions.[24]

- The modification of functions, operations, and input-output actions do not have to be performed separately for each object class but, as the user wishes, can cover all object classes needed to perform a specific task. In this way, generic functions are not tied to a particular application, but are made independent and cross-applicational in modularized form. Special functions are assembled for particular tasks.

- This approach facilitates a substantial reduction in system complexity but, in so doing, does not restrict system performance, as is the case with today's concepts.

- It allows a far more flexible response by the technical support system to the natural dynamics of the organizational environment (especially task tailoring and combinations of sector-targeted work actions).

- It thereby meets a very central demand for a genuinely human-centred technological design, which has been a long time in discussion but has not yet been fulfilled.

- It can contribute to overcoming problems of cooperation that stand in the way of user-friendly system adaptation in many organizations today.

---

[24] Most of the systems currently available try to achieve integration by adding on so-called "rucksacks". However, these solutions substantially increase complexity and are ultimately inefficient in comparison with the special systems.

References

Anderson, J. R. (1983):
The architecture of cognition. Cambridge, MA: Harvard University Press.

Bauer, J., Herberg, V.D., & Schwab, Th. (1987):
Hilfesysteme. In: K. P. Fähnrich (Ed.), *Software-Ergonomie. State of the Art.* München: Oldenbourg, pp. 118–128.

Benyon, D., Innocent, P. & Murray, D. (1987):
System adaptivity and the modelling of stereotypes. In: H.J. Bullinger, & B. Shackel (Eds), *Proceedings of the Conference Human-Computer Interaction - INTERACT '87.*

Benyon, D. & Murray, D. (1988): Experience with adaptive interfaces. *The Computer Journal,* 31, 465–473.

Budde, R., Kautz, K., Kuhlenkamp, K., & Züllighoven, H. (1992):
Prototyping: An approach to evolutionary system development. Berlin: Springer.

Carroll, J.M., & Carrithers, C. (1984):
Training wheels in a user interface. *Communications of the ACM, 27,* 800–806.

DIN; (1988, February):
Bildschirmarbeitsplätze: Grundsätze ergonomischer Dialoggestaltung. Berlin: Beuth.

Döbele-Berger, C., Schwellach, G., van Treek, W. & Zimmer, G. (1988):
Softwarenutzung am Arbeitsplatz und berufliche Weiterbildung. Eine explorative Studie. Arbeitspapier der Forschungsgruppe Verwaltungsautomation 47. Kassel.

Eberleh, E. (1989):
Beschreibung, Klassifikation und mentale Repräsentation komplexer Mensch–Computer-Interaktion. Regensburg: Roderer.

Falck, M. (1991):
Arbeit in der Organisation. Zur Rolle der Kommunikation als Arbeit in der Arbeit und als Gegenstand technischer Gestaltung (working–paper). Berlin.

Floyd, C., Züllighoven, H., Budde, R., & Keil-Slawik, R. (Eds.).(1992):
Software development and reality construction. Berlin: Springer.

Foerster, H. v. (1984):
Principles of self-organization: In a socio–managerial context. In: H. Ulrich, G. Probst (Eds.), *Self-organization and management of social systems. Insides, promises, doubts, and questions.* Heidelberg: Springer, pp. 2–25.

Foerster, H. v.; Floyd, C. (1992):
Self-organization and software development. In: C. Floyd, H. Züllighoven, R. Budde, R. Keil-Slawik (Eds.), *Software development and reality construction*. Berlin: Springer, pp. 75–85.

French, W.L., & Bell, C.H.Jr. (1973):
Organization development. Englewood Cliffs, N. J. : Prentice-Hall.

Friedrich, J. (1990):
Adaptivität und Adaptierbarkeit informationstechnischer Systeme in der Arbeitswelt - zur Sozialverträglichkeit zweier Paradigmen. In: A. Reuter, (Ed.), *Informatik auf dem Weg zum Anwender*. 20. GI - Jahrestagung. Berlin, pp. 178–191.

Glasersfeld, E. v. (1987):
Wissen, Sprache und Wirklichkeit. Arbeiten zum radikalen Konstruktivismus. Braunschweig: Vieweg.

Gottschall, K., Mickler, O., & Neubert, J. (1985):
Computerunterstützte Verwaltung. Auswirkungen der Reorganisation von Routinearbeiten. Schriftenreihe "Humanisierung des Arbeitslebens", Band 60. Frankfurt/M.: Campus.

Hacker, W. (1978):
Allgemeine Arbeit- und Ingenieurpsychologie: Psychische Struktur und Regulation von Arbeitstätigkeiten (2. überarbeitete Aufl.), Bern: Huber.

Hacker, W. (1987):
Software-Gestaltung als Arbeitsgestaltung. In: K.P. Fähnrich (Ed.), *Software-Ergonomie*. München: Oldenbourg, pp. 29–42.

Haken, H. (1991):
Erfolgsgeheimnisse der Natur. Synergetik, die Lehre vom Zusammenwirken. Frankfurt/M.: Ullstein.

ISO 9241, Part 10:
Ergonomic requirements for office work with visual display terminals (VDTs): Dialogue principles.

Johnson-Laird, P. N. (1983):
Mental models. Cambridge, Great Britain: Cambridge University Press.

Karger, C. (1990):
Klassifikation von Benutzermodellen in adaptiven Systemen (GMD–working–paper).

Karger, C., & Oppermann, R. (1991):
Empirische Nutzungsuntersuchung adaptierbarer Schnittstelleneigenschaften. In: D. Ackermann, & E. Ulich (Eds.), *Software-Ergonomie '91: Benutzerorientierte Software-Entwicklung*. Stuttgart: Teubner, pp. 272–280.

Kieser, A., & Kubicek, H. (1978):
Organisationstheorien. Stuttgart: Kohlhammer.

Kubicek, H. (1986)
Konzeptionelle Herausforderung bei der sozialverträglichen Gestaltung der sogenannten Neuen Informations- und Kommunikationstechniken. In: A. Rolf (Ed.), *Neue Techniken Alternativ.* Hamburg: VSA, pp. 81–98.

Lehmann, M.M. (1982):
Program evolution (Research Report Doc 82/1). London: Imperial College of Science and Technology, Department of Computing.

Luhmann, N. (1987):
Soziale Systeme: Grundriß einer allgemeinen Theorie. Frankfurt/M.: Suhrkamp.

Luhmann, N. (1991):
Soziologie des Risikos. Berlin: de Gruyter.

Maturana, H.; & Varela, F. (1982):
Autopoietische Systeme: eine Bestimmung der lebendigen Organisation. In: H. Maturana (Ed.), *Erkennen: Die Organisation und Verkörperung von Wirklichkeit. Ausgewählte Arbeiten zur biologischen Epistemologie* (2nd ed.). Braunschweig: Vieweg, pp. 170–235.

Maturana, H. and Varela, F. (1990):
El arbol del conocimiento: Las bases biólogicas del conocimiento humano. Madrid: Editorial Debate.

Minsky, M. (1981):
Framework for representing knowledge in mind design. In: R. Brachman, & H. Levesque (Eds.), *Readings in knowledge representation.* Los Altos, CA: Morgan Kaufmann, pp. 245–262.

Nake, F. (1986):
Die Verdopplung des Werkzeugs. In: A. Rolf (Ed.), *Neue Techniken alternativ.* Hamburg: VSA, pp. 43–52.

Nippa, M. (1988):
Gestaltungsgrundsätze für die Büroorganisation: Konzepte für eine informationsorientierte Unternehmensentwicklung unter Berücksichtigung neuer Bürokommunikationstechniken. Berlin: Schmidt.

Norman, D. A., & Rumelhart, D. E. (1975):
Explorations in Cognition. San Francisco: Freeman.

Oberquelle, H. (Ed.) (1991):
Kooperative Arbeit und Computerunterstützung: Stand und Perspektiven. Göttingen: Verlag für Angewandte Psychologie.

Oppermann, R., Murchner, B., Paetau, M., Pieper, M., Simm, H. & Stellmacher, I. (1989):
Evaluation of dialog systems (GMD-Studien Nr. 169). Sankt Augustin: GMD-Selbstverlag.

Paetau, M. (1991):
Mensch–Maschine-Kommunikation im Spannungsfeld zwischen Individualisierung und Standardisierung (GMD-Arbeitspapier 520). Sankt Augustin: GMD-Selbstverlag.

Paetau, M. (1990):
Mensch-Maschine-Kommunikation: Software, Gestaltungspotentiale, Sozialverträglichkeit. Frankfurt/M.: Campus.

Picot, A., & Reichwald, R. (1984):
Bürokommunikation, Leitsätze für den Anwender. München: CW-Publikationen.

Riegas, V., & Vetter, Ch. (1990):
Gespräch mit Humberto Maturana. In: V. Riegas, & Ch. Vetter (Eds.), *Zur Biologie der Kognition.* Frankfurt/M.: Suhrkamp, pp. 11–90.

Roth, G. (1987):
Autopoiese und Kognition: Die Theorie H.R. Maturanas und die Notwendigkeit ihrer Weiterentwicklung. In: S.J. Schmidt (Ed.), *Der Diskurs des radikalen Konstruktivismus.* Frankfurt/M.: Suhrkamp, pp. 256–286.

Schmid, M. (1987):
Autopoiesis und soziales System: Eine Standortbestimmung. In: H. Haferkamp, & M. Schmid (Eds.), *Sinn, Kommunikation und soziale Differenzierung: Beiträge zu Luhmanns Theorie sozialer Systeme.* Frankfurt/M.: Suhrkamp, pp. 25–50.

Simm, H. (1991):
Adaptierbarkeit in marktgängigen Systemen (GMD-Arbeitspapier 504). Sankt Augustin: GMD-Selbstverlag.

Smieja, F., & Mühlenbein, H. (1991):
Von und mit der Evolution lernen. GMD-Spiegel 2/91, 32–35.

Szyperski, N., Grochla, E., Höring, K. , & Schmitz, P. (1982):
Bürosystem in der Entwicklung: Studie zur Typologie und Gestaltung von Büroarbeitsplätzen. Braunschweig: Vieweg.

Türk, K. (1989):
Neuere Entwicklungen in der Organisationsforschung. Stuttgart: Enke.

Ulich, E. (1978):
Über das Prinzip der differentiellen Arbeitsgestaltung. Industrielle Organisation, 47, 566–568.

VDI-Richtlinie 5005 (1990):
Software-Ergonomie in der Bürokommunikation. Düsseldorf.

Weber, M. (1972):
Wirtschaft und Gesellschaft. (5th ed.). Tübingen: Mohr.

Weltz, F. (1986):
Wer wird Herr der Systeme? Der Einsatz neuer Bürotechnologie und die innerbetriebliche Handlungskonstellation. In: R. Seltz, U. Mill, & E. Hildebrandt (Eds.), *Organisation als soziales System.* Berlin: Sigma, pp. 151–161.

Wingert, B. and Riehm, U. (1985):
Computer als Werkzeug. Anmerkungen zu einem verbreiteten Mißverständnis. In: *Technik und Gesellschaft* (Jahrbuch 3). Frankfurt/M.: Campus, pp. 107–131.

Winograd, T. and Flores, F. (1986):
Understanding computers and cognition: A new foundation for design. Norwood, NJ: Ablex.

Womack, J. P., Jones, D. T., Roos, D. (1990):
The machine that changed the world. New York: Rawson.

Subject Index

Author Index

Printed and bound by CPI Group (UK) Ltd, Croydon, CR0 4YY

17/10/2024

01775684-0009